E. Sherman Gould

The Elements of water supply Engineering

E. Sherman Gould

The Elements of water supply Engineering

ISBN/EAN: 9783743343450

Manufactured in Europe, USA, Canada, Australia, Japa

Cover: Foto ©ninafisch / pixelio.de

Manufactured and distributed by brebook publishing software (www.brebook.com)

E. Sherman Gould

The Elements of water supply Engineering

THE ELEMENTS

OF

WATER SUPPLY

ENGINEERING.

BY

E. SHERMAN GOULD,
M. AM. SOC. C. E.

NEW YORK:
THE ENGINEERING NEWS PUBLISHING CO.
1899.

COPYRIGHTED, 1899,
BY THE ENGINEERING NEWS PUBLISHING CO.

78630.

PREFACE.

"Practical Hydraulic Formulæ" was first published in 1889. The second edition appeared in 1894 in a form much extended by notes on THE QUALITY and THE QUANTITY OF WATER and on THE CALCULATION and THE CONSTRUCTION OF DAMS. In the present work all the matter contained in the first and second editions of "Practical Hydraulic Formulæ" is republished; but, under the heading of "NOTES TO PARTS I. AND II.," it has been supplemented with copious memoranda, elucidating and greatly extending many important points inadequately treated in the previous editions.

An entirely new part (PART III.) has been added. This treats of the FLOW OF WATER THROUGH MASONRY CONDUITS on the basis of Darcy's formulæ, with some practical details of aqueduct and tunnel construction. A few paragraphs are devoted to the subject of the FILTRATION OF PUBLIC WATER SUPPLIES, sufficient, it is hoped, to indicate the proper lines of further investigation to those who are interested in pursuing them. A somewhat fuller treatment is given to the subject of PUMPING ENGINES AND DUTY TRIALS. Some pages have been added on the subject of ARCHES AND ABUTMENTS. As an extended project of water supply frequently embraces the construction of arched masonry aqueducts, it is believed that the brief and practical rules given in this part of the book will prove acceptable to the hydraulic engineer. A set of carefully calculated, labor-saving TABLES, with explanations and examples, terminates the book.

It will be seen that the present work covers so wide a field that to retain for it as a whole the title originally given to the first part would be misleading. It is, therefore, called "THE ELEMENTS OF WATER SUPPLY ENGINEERING," which name more truly indicates its scope.

While it is believed that every topic connected with water-supply engineering has been at least touched upon, especial pains have been taken to go into very close detail in the matter of the principal *dimensions* and *quantities* involved in the designing of hydraulic work. As these are the points which the author has most carefully sought for in his own reading and observation, so he believes that they are the ones which others may be most interested in finding fully treated of in the present volume.

E. SHERMAN GOULD.

TABLE OF CONTENTS.

INTRODUCTION.

CHAPTER I.

Flow through a short horizontal pipe—Effect on velocity of increased length—Frictional head—Hydraulic grade line—Hydrostratic and hydraulic pressures—Piezometric tubes—Results of raising a pipe line above the hydraulic grade line—Why the water ceases to rise in the upper stories of the houses of a town when the consumption is increased—Influence of inside surface of pipes upon velocity of flow—Darcy's coefficients—Fundamental equations—Length of a pipe line usually determined by its horizontal projection—Numerical examples of simple and compound systems - - - - - - - - Pages 11-24

CHAPTER II.

Calculations are the same for pipes laid horizontally or on a slope—Qualification of this statement—Pipe of uniform diameter equivalent to compound system—General formula—Numerical example—Use of logarithms (foot note)—Numerical example of branch pipe—Simplified method—Numerical examples—Relative discharges through branches variously placed—Discharges determined by plotting—Caution regarding results obtained by calculation—Numerical examples - - Pages 25-38

CHAPTER III.

Numerical example of a system of pipes for the supply of a town—Establishment of additional formulæ for facilitating such calculations—Determination of diameters—Pumping and reservoirs—Caution regarding calculated results—Useful approximate formulæ—Table of 5th powers—Preponderating influence of diameter over grade illustrated by example - - - - - - - - - - Pages 39-48

CHAPTER IV.

Use of formula 14 illustrated by numerical example of compound system combined with branches—Comparison of results—Rough and smooth pipes—Pipes communicating with three reservoirs—Numerical examples under varying conditions—Loss of head from other causes than friction—Velocity, entrance and exit head—Numerical examples and general formula—Downward discharge through a vertical pipe—Other minor losses of head—Abrupt changes of diameter—Partially opened valve—Branches and bends—Centrifugal force—Small importance of all losses of head except frictional in the case of long pipes—All such covered by "even inches" in the diameter . Pages 49–63

CHAPTER V.

    Notes on pipelaying - - - - - - - - Pages 64–67

APPENDIX.

    Weight of cast iron pipe—Various useful formulæ . Pages 68–70

SECOND PART.

NOTES ON WATER SUPPLY ENGINEERING.

    Quality of water—Quantity of water—Dams, calculation of; construction of—Reference to other publications - - Pages 71-105

    Notes to Parts I and II - - - - - - - Pages 106-122

THIRD PART.

Flow of Water Through Masonry Conduits - -	Pages 123-125
Some Details of Tunnel and Acqueduct Construction	Pages 125-126
Filtration of Public Water Supplies - - - -	Pages 127-129
Pumping Engines and Duty Trials - - - -	Pages 129-137
Arches and Abutments - - - - - - -	Pages 137-154
Hydraulic Tables - - - - - - -	Pages 154-162
Index - - - - - - - - - -	Pages 163-168

INTRODUCTION TO HYDRAULIC FORMULÆ.

The following pages first appeared as a series of articles in the columns of ENGINEERING NEWS. They are now republished with a few corrections and additions.

In virtue of the law of gravitation, water tends naturally to pass from a higher to a lower level, and without a difference of level there can be no natural flow.

It can be said in all seriousness—although the statement may seem to invite the unjust accusation of an ill-timed attempt at pleasantry—that the whole science of hydraulics is founded upon the three following homely and unassailable axioms:

First. That water always seeks its own lowest level.

Second. That, therefore, it always tends to run down hill, and

Third. That, other things being equal, the steeper the hill, the faster it runs.

In the case of water flowing through long pipes, the hill down which it tends to run is the HYDRAULIC GRADE LINE. If the pipe be of uniform diameter and character, the hydraulic grade line is a straight line joining the water surfaces at its two extremities, provided that the pipe lies wholly below such straight line, and its declivity is measured—like that of all hills —by the ratio of its height to its length.

But if there be any changes whatever in the pipe, either in diameter or in the nature of its inside surface; or if there be increase or diminution of the volume of water entering it at its upper extremity by reason of branches leading to or from the main pipe, then the hydraulic grade line becomes broken and distorted to a greater or less extent, so that its declivity is not uniform from end to end, but consists of a series of varying grades some steeper than others though all sloping in the same direction.

As regards the third axiom, the proviso—" other things being equal"—must not be overlooked. For we shall find that a pipe of greater diameter but less hydraulic declivity than another, may give a greater velocity to the water passing through it. Also, of two pipes of the same hydraulic slope and diameter, the one having the smoother inside surface affords the greater velocity.

The vertical distance from any point in a pipe to the hydraulic grade line, constitutes the *Piezometric height*, and measures the hydraulic pressure at that point. It will be seen that the solution of problems relating to the flow of water through pipes, lies in the knowing or ascertaining of the piezometric height at any desired point. In general, it is necessary to establish the piezometric height for every point of change of any kind which occurs throughout the entire length of the conduit. The joining of the upper extremities of these heights gives the complete hydraulic grade line.

The object of the following papers is to establish systematic methods for tracing the hydraulic grade line under the different circumstances likely to occur in practice, and generally, to furnish solutions for a large number of practical problems, commencing with the simplest cases and extending to some rather intricate ones, not usually embraced in our hydraulic manuals.

E. S. G.

SCRANTON, Pa., May, 1889.

HYDRAULIC FORMULÆ.

CHAPTER I.

Flow through a Short Horizontal Pipe—Effect on Velocity of Incr.ased Length—Frictional Head—Hydraulic Grade Line -Hydrostatic and Hydraulic Pressures—Piezometric Tubes—Result of Raising a Pipe Line Above the Hydraulic Grade Line—Why the Water Ceases to Rise in the Upper Stories of the Houses of a Town when the Consumption is Increased—Influence of Inside Surface of Pipes Upon Velocity of Flow—Darcy's Coefficients—Fundamental Equations—Length of a Pipe Line Usually Determined by its Horizontal Projection—Numerical Examples of Simple and Compound Systems.

Let us suppose a reservoir of large relative area and depth to be tapped near its bottom by a horizontal cylindrical pipe, of which the length is equal to about three times its diameter.

If there were no physical resistance to the flow, the velocity of the water issuing from the pipe would be given by the formula for the velocity of falling bodies :

$$V = \sqrt{2gH} = 8.02 \sqrt{H},$$

in which $V =$ velocity in feet per second, $g =$ the acceleration due to gravity $= 32.2$ ft., and $H =$ the height, expressed in feet, of the surface of the water in the reservoir above the center of the pipe.

Observation shows, however, that in the case cited the velocity of discharge is equal only to that theoretically due to a height of about two-thirds of H; that is :

$$V = \sqrt{\frac{4gH}{3}} = 6.55 \sqrt{H}.$$

The remaining third of the height is consumed in overcoming the resistance offered to entry by the edges of the orifice to the inflowing vein of water. The head necessary to overcome the resistance to entry is therefore about one-half of that necessary to produce the velocity of flow.

If the length of the pipe should be increased progressively and indefinitely, the velocity would be found to diminish inversely as the square root of the length. It would correspond, therefore, to a smaller and smaller percentage of the total head H. The resistance to entry diminishes directly as the velocity, and the head necessary to overcome it is always equal to about one-half of that necessary to produce the given velocity as calculated by the laws of falling bodies.

As the length of the pipe (always supposed to remain horizontal) increases, and the velocity of discharge diminishes, the sum of these two heads, *i. e.*, one and a half times that necessary to produce the actual velocity, is no longer equal to the total head H, as we have seen to be the case when the length of the pipe is only about three diameters. What, then, becomes of the remainder of H? It is consumed in overcoming the increasing frictional resistances engendered by contact of the moving water with the inside surface of the pipe. When the pipe is very long, and the velocity therefore relatively low, the sum of the velocity and entrance heads is small, and by far the greater part of the total head is required to force the water through the pipe against the opposition offered by friction to its flow. In such cases, which are those occurring most generally in practice when water is conveyed from a reservoir for the supply of a town, the velocity and entrance heads are commonly ignored, and the total head H is supposed to be available for overcoming the frictional resistances. As this occasions, however, an error—although generally a very small one—in the *wrong direction*, judgment is required in exercising this latitude. Later on we will revert to this point, but for the present we will consider only frictional resistances, particularly since—and indeed because—in practice our assumed data are almost always sufficient to afford an ample margin to cover the neglected factors.

In what precedes we have considered a horizontal pipe issuing from a reservoir in which the surface of the water is maintained at a constant level. In practice these conditions rarely obtain.

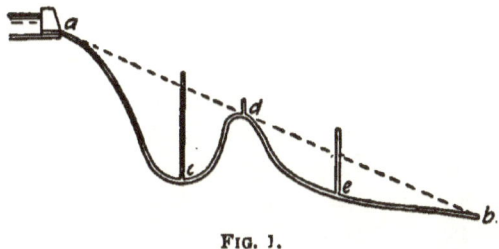

Fig. 1.

Suppose a system, such as is shown by Fig. 1, consisting of a reservoir and pipe line of varying and contrary slopes. As the level of the water in the reservoir would be subject to fluctuations, and liable at times to be greatly drawn down, it is customary to consider the surface of the water as standing at its lowest possible level, *i. e.*, the mouth of the pipe. In this case, the value of H would be equal to the difference of level of the two extremities a and b of the pipe, and the line ab joining the centers of the two ends would form what is called *the hydraulic grade line*, the establishing of which is the first step to be taken in laying out a system of water supply.

Suppose that at the points c, d, and e vertical tubes, open at their upper ends, were connected with the pipe. The water, when flowing freely from the end b of the pipe would rise in each of these tubes to about the height of the hydraulic grade line at these points, and, if branches were connected at the points c, d, and e, they would, when closed, sustain a pressure upon their gates equal to the head comprised between the gates and the grade line. If the gates were open, the branches would discharge water under heads equal to the difference of level of the hydraulic grade line at the point of embranchment and their remote extremities, less a certain amount depending upon the volume discharged, which will be spoken of hereafter.

At d, where the top of the pipe just touches the grade line, there would be no pressure at all when the water was flowing through the pipe, except the very small amount due to the depth of water in the pipe itself.

If the end b should be closed so that there was no movement of water in the pipe, the water would rise in the tubes, if they were long enough, until it stood at the same level as the water in the reservoir, and the pressures at c, d, and e would be equal to the head comprised between these points and the level of the water in the reservoir. The latter is called the *hydrostatic pressure*, or simply the *static pressure*, and the former the *hydraulic pressure*, at these points.

The tubes spoken of are known by the name of *piezometric tubes*.

The importance of correctly establishing the hydraulic grade line is illustrated by reference to a case such as is shown in Fig. 2, in which the pipe, at the point c, rises above the grade line ab. To explain: It will be readily deduced from what has been already said in reference to horizontal pipes that the velocity of flow, and consequently the delivery, of a pipe increases with the steepness of its slope. In this case the pipe ab is divided into two parts, the one ac with a hydraulic grade line flatter than ab, and the other cb with one steeper than ab. The delivery of the entire system, if the pipe were of the same diameter throughout, would be governed by the flatter portion ac, and the portion cb would be capable, in virtue of its steeper slope, of discharging a greater volume of water than it could receive from ac. Consequently it would act merely as a trough and would never run full, and if a piezometric tube were placed in it at d for instance, no water would rise in the tube, and no pressure be exerted.

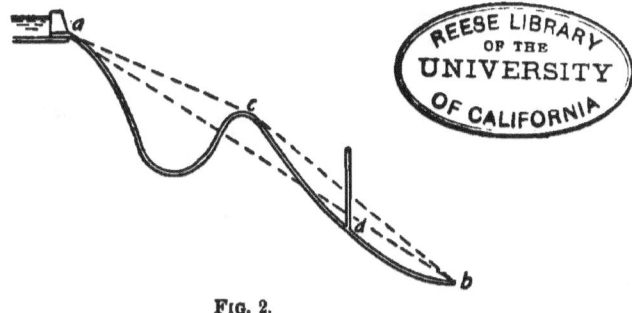

Fig. 2.

It is very important, therefore, in locating a pipe line that the pipe should nowhere rise above the hydraulic grade line. The full amount of water could indeed be carried over the high point c by means of siphonage, but this expedient is not resorted to in practice. Should the nature of the ground require such a location as that shown in Fig. 2, it would be necessary to increase the diameter of the pipe between a and c, so that it would deliver the required volume under the reduced head, and to diminish that between c and b, so that it should only deliver the same volume under its increased head, and therefore run full. The calculations necessary to determine the proper diameters will be shortly developed.

Should the axis of the pipe coincide exactly with the hydraulic grade line ab, the pipe would run full (provided the feed were sufficient) but would be under no pressure, and no water would rise in piezometric tubes placed on any part of the pipe. Moreover, as the slope would be the same for any portion of the pipe, the velocity and delivery would be unchanged, whether we cut the pipe off at a comparatively short length, or extend it indefinitely.

As a further and very interesting practical illustration of the effects of a hydraulic grade line of varying steepness, let us con-

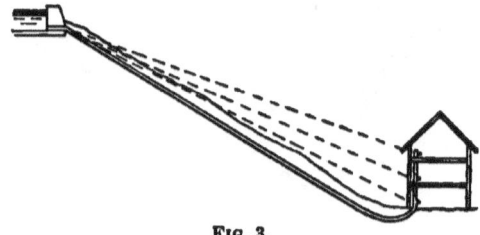

FIG. 3.

sider (Fig. 3) the case of a house supplied with water by a pipe communicating with a reservoir.

Suppose the pipe to be sufficiently large to furnish a certain volume of water per hour to the upper story of the house. If now a larger volume were required, it is clear that, unless we increase the diameter of the pipe, it would be necessary to increase the steepness of pitch of the grade line, in other words, to increase the head, or difference of level between the reservoir and the point of discharge. The increased volume could therefore be only drawn from a lower story.*

Or, to put the same conditions under a different form, suppose, as before, the pipe to be just large enough to supply the top story of the house, the taps on the lower floors being closed. Should they be opened, it is evident that a greater amount of water would be discharged from them than from the upper one, because they would discharge under a greater head. The result would be a diminished flow or perhaps no flow at all on the top floor, and an increased discharge of water at a lower level.

---

* In other words, if we wish to increase the *volume*, the diameter of pipe remaining constant, we must increase the *velocity;* and the increased velocity can only be obtained by an increased difference of level between the two ends of the pipe. If the elevation of the upper end, or surface of water in the reservoir, cannot be increased, that of the lower end, or point of discharge, must be diminished.

PRACTICAL HYDRAULIC FORMULÆ.   17

This case shows why the water ceases to rise in the upper stories of the houses of a town when the consumption increases.

It has been found by observation that the velocity of water flowing through pipes is greatly affected by the nature of their inside surface, increasing with the smoothness and diminishing with the roughness of the same. By direct experiment, coefficients have been established for different conditions of surface. It has also been found that these coefficients vary slightly with the diameter of the pipe, a pipe of a certain size giving a greater velocity than one of the same character of inside surface but of smaller diameter, the differences becoming smaller as the diameters increase.

The value of this coefficient, which will be designated throughout this paper by C, is given below for a number of different diameters and for two classes of pipes,—those which are clean and smooth on the inside, and those which are rough and incrusted, the difference being as 2 to 1. As all pipes, after a few years of service, are liable to become more or less roughened and obstructed by deposits, it is always safer when calculating the proper diameters of a permanent water supply, to assume rough pipes at once, although diameters thus calculated will, for perhaps a number of years, deliver quantities greatly in excess of the desired amounts.

The coefficients given below are those determined experimentally by DARCY. Of course, in the subsequent calculations which will be made, any other values might be substituted for the ones given. It is well to remark, however, in regard to the coefficient, that although this factor is a controlling one in the calculation of the discharge of pipes, it is useless to attempt an excessive refinement in establishing its value, because not only is it difficult to determine this value with exactness for a given diameter and condition of pipe, but this condition, and even the diameter of the pipe, is liable to undergo considerable variation in the same pipe in the course of a few years.

TABLE OF COEFFICIENTS.

Diameter in inches.	Value of C for rough pipes.	Value of C for smooth pipes.
3	0.00080	0.00040
4	0.00076	0.00038
6	0.00072	0.00036
8	0.00068	0.00034
10	0.00065	0.00033
12	0.00066	0.00033
14	0.00065	0.000325
16	0.00064	0.00032
24	0.00064	0.00032
30	0.00063	0.000315
36	0.00062	0.00031
48	0.00062	0.00031

In all the following calculations, the coefficient for rough pipes will be used.

The two fundamental equations relating to the flow of water through long pipes are:

$$\frac{D \times H}{L} = CV^2 \quad (1)$$

$$Q = AV \quad (2)$$

Equation No. 2 will generally be written:

$$Q = A \sqrt{\frac{D \times H}{C \times L}} \quad (3)$$

by taking the value of V from (1).

The first of these has been established by DARCY; the second is based upon a self-evident proposition.

In these equations:

D = diameter of pipe in feet
H = total head " "
L = length of pipe " "
C = coefficient
V = mean velocity in feet per second
Q = discharge in cubic feet per second
A = area of pipe in square feet = $D^2 \times 0.785$

The above two formulæ solve, directly or indirectly, all prob-

lems relating to the flow through long pipes, and all such problems must be brought into a form admitting of their application, in order to obtain a solution.

It will be observed that $\dfrac{H}{L}$ is the rise or fall per foot of length of pipe, and is therefore the natural sine of the inclination of the slope to the horizon. This relation is frequently used under the form $I = \dfrac{H}{L}$. Using this notation, (1) would be written:

$$DI = C V^2$$

In long pipes the length is generally taken as being equal to the horizontal distance separating the two ends of the pipe, as the difference between this distance and the actual length of the pipe is relatively insignificant. If, however, a case should present itself in which this difference was considerable, the actual length of pipe should be taken. Further on, an extreme case of this kind will be given, presenting some interesting features.

Some practical examples of the use of these formulæ will now be given. In all that follows, the resistances of entry, exit, and velocity will be neglected, and the total head will be considered as available for overcoming friction. The examination of cases where the above factors are included is reserved for a later portion of this paper, as they are of secondary importance when dealing with long pipes.

Example 1.—A pipe, 1 ft. in diameter and 1,000 ft. long, has a total fall of 10 ft. What are the velocity and volume of its discharge?

Substituting the given values in (1) we have:

$$\dfrac{1 \times 10}{1000} = 0.00066\ V^2$$

$$V = 3.39 \text{ ft. per second.}$$

Using this value of V in (2), we have :

$$Q = 0.785 \times 3.89$$
$$Q = 3.055 \text{ cu. ft. per second.}$$

Example 2.—Two reservoirs, having a difference of level of water surface of 30 ft., are joined by a pipe 3,000 ft. long. What should be the diameter of the pipe to deliver 16 cu. ft. of water per second from the upper to the lower reservoir ?

Eliminating V between (1) and (2) we have :

$$\frac{D \times H}{L \times C} = \frac{Q^2}{A^2}.$$

Observing that $A = D^2 \, 0.785$;

$$\frac{D \times H}{L \times C} = \frac{Q^2}{D^4 \times 0.616}$$

Whence

$$D^5 = \frac{Q^2 \times L \times C}{H \times 0.616} \qquad (4)$$

If we knew the proper value of the coefficient C in the above equation, it could be immediately solved, and the value of D obtained. But C varies with the diameter, and the diameter is as yet unknown. We must therefore have recourse to "Trial and Error" for a solution.

Suppose it should appear to us, at first sight, that a 12-in. pipe was likely to be of the proper size. We therefore take $C = 0.00066$, and write :

$$D^5 = \frac{256 \times 3000 \times 0.00066}{30 \times 0.616}$$
$$D^5 = 27.70$$
$$D = 1.94 \text{ ft.}$$

From this we see that the pipe should be nearly 2 ft. in diameter, and as we have taken too large a coefficient (that for 24 ins. = 0.00064), we are sure that 1.94 is too large. As pipes are never made of fractional diameters, the above value of D would be taken = 24 ins., and therefore we would push the calculation no further. If the case had happened to be one requiring minute accuracy, we would recalculate the above equation, using 0.00064 for the value of C. The result would be, $D = 1.93$ ft. nearly, practically the same as the value already obtained.

The above examples (which are those commonly occurring in practice) are very simple, and involve only the direct application of the fundamental formulæ. Let us now consider cases of a more complicated character, where they can only be used indirectly, and where a certain amount of judgment and tact is required in the preparation of the data.

Example 3.—Suppose a reservoir R (Fig. 4) containing a depth of water of 50 ft. above the center of the horizontal pipe A, 1 ft. in diameter and 1,000 ft. long, connected by a reducer with another horizontal pipe B, 2 ft. in diameter and 3,000 ft. long. It is required to calculate the piezometric head h at the junction, from which the discharge can be calculated, and the hydraulic grade line abc established.

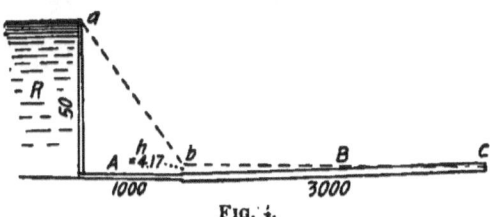

FIG. 4.

It is evident that the 24-in. pipe must, under the head h, discharge the same quantity per second as the 12-in. pipe, under the head $50 - h$. We have then from (3) the equality:

$$3.14 \sqrt{\frac{2 \times h}{3000 \times 0.00064}} = 0.785 \sqrt{\frac{1 \times (50 - h)}{1000 \times 0.00066}}.$$

Dividing by 0.785, squaring, and simplifying:

$$\frac{h}{0.02} = \frac{50 - h}{0.22},$$

whence
$$h = 4.17.$$

We can now very readily get the discharge, by substituting

the value 4.17 for h in either member of the above equality. Thus:

$$Q = 3.14 \sqrt{\frac{4.17}{0.96}} = 6.54 \text{ cu. ft. per sec.}$$

Verifying in the other member—a precaution which should never be neglected—we obtain the same result.

It is evident that the diameter of B may be assumed so large that no value of h can be found to satisfy the condition that both pipes shall run full with the given height of water in the reservoir. In such a case the pipe B serves only as a trough to receive the water discharged through A under a head of 50 ft.

Suppose that in the above example the places of the two pipes, A and B, should be changed. Evidently we should have:

$$h = 45.83.$$

This piezometric height would give, with the transposed position of the pipes, the same discharge as before, the only difference being a notable change in the hydraulic grade line. If the pipes were tapped by branches, the greater elevation of the grade line in this case would bring a much greater pressure upon the branches, enabling them to deliver water at a higher level than in the first position of the pipes.

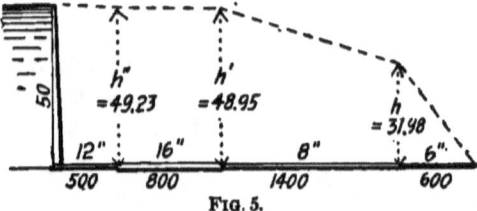

FIG. 5.

The above example may be extended so as to cover cases where pipes of several different diameters are used. Thus, suppose a system of pipes, such as is shown in Fig. 5, where a reservoir with a head of 50 ft. of water, as before, is tapped by a horizontal line of pipes, consisting in order of 500 ft. of 12-in.; 800 ft. of 16-in., 1,400 ft. of 8-in., and 600 ft. of 6-in. pipe.

PRACTICAL HYDRAULIC FORMULÆ. 23

This example may be worked in the same way as the previous one, by getting equations for h, h', and h'' expressed, by substitution, in terms of h. But it will be easier to treat the question in another way, which will also exhibit the further resources which we have at our disposal in solving hydraulic problems.

Since each section of pipe must discharge equal volumes in equal times, it is evident that the respective velocities of flow must vary inversely as the areas of the pipes. These areas vary as the squares of the different diameters. Designating, therefore, by V the lowest rate of velocity, $i.\ e.$, that of the water passing through the largest pipe (the 16-in. one), we obtain the relative velocities in the other pipes by multiplying V by the ratio of the square of the diameter of the largest pipe to the squares of the other diameters. It will be convenient to form the following table:

Lengths in ft.	Diameters in ft.	Velocities in ft. per second.
500	1	1.78 V
800	1¼	V
1,400	⅞	4 V
600	½	7.11 V

Beginning at the lower end of the system, that is with the 6-in. pipe, and employing formula (1) in which h and V are the unknown quantities, we have:

$$\frac{1}{2} \times \frac{h}{600} = 0.00072 \times (7.11)^2 \times V^2;$$

whence: $h = 43.68\ V^2$

again: $\dfrac{2}{3} \times \dfrac{(h' - h)}{1400} = \dfrac{2}{3} \times \left(\dfrac{h' - 43.68\ V^2}{1400}\right) = 0.00069 \times (4)^2 \times V^2;$

whence: $h' = 66.86\ V^2$

similarly: $\dfrac{4}{3} \times \dfrac{(h'' - h')}{800} = \dfrac{4}{3} \times \left(\dfrac{h'' - 66.86\ V^2}{800}\right) = 0.00065 \times V^2$

whence: $h'' = 67.25\ V^2$

Finally: $\dfrac{50 - h''}{500} = \dfrac{50 - 67.25\ V^2}{500} = 0.00066 \times (1.78)^2 \times V^2$

whence: $V^2 = 0.7321$
$V = 0.8556$ ft. per second.

Substituting this value of V^2 in the above equations:

$$h = 31.98 \text{ ft.}$$
$$h' = 18.95 \text{ "}$$
$$h'' = 49.23 \text{ "}$$

We also get the velocities in the different pipes, thus:

6 inch, velocity = $7.11 \times 0.856 = 6.086$
8   "   "   = $4 \times 0.856 = 3.424$
16 "   "   = $1 \times 0.856 = 0.856$
12 "   "   = $1.78 \times 0.856 = 1.524$

The work can be checked by using the above values of h, h', and h'', along with the other data, in (1), and obtaining the velocities in this way.

Thus, beginning with the 6-in. pipe:

$$\frac{1}{2} \times \frac{31.98}{600} = 0.00072 \; V^2$$
$$V = 6.08$$
$$\frac{2}{3} \times \frac{16.97}{1400} = 0.00069 \; V'^2$$
$$V' = 3.42$$
$$\frac{4}{3} \times \frac{0.28}{800} = 0.00065 \; V''^2$$
$$V'' = 0.85$$
$$1 \times \frac{0.77}{500} = 0.00065 \; V'''^2$$
$$V''' = 1.53$$

A very close agreement throughout.

In the above calculations the decimals have been carried out further than would ordinarily be necessary in practice. It was done in the present instance in order to avoid discrepancies in checking.

We have another check, in the volumes discharged. Thus the discharge through the 6-in. pipe, with the given velocity, is by (2):

$$Q = 0.195 \times 6.086$$
$$Q = 1.19 \text{ cubic ft. per second.}$$

All the other pipes should have an equal discharge; for instance, the 12-in. pipe gives:

$$Q = 0.78 \times 1.524$$
$$Q = 1.19 \text{ cubic ft. per second.}$$

CHAPTER II.

Calculations are the Same for Pipes laid Horizontally or on a Slope—Qualification of this Statement—Pipe of Uniform Diameter Equivalent to Compound System —General Formula—Numerical Example—Use of Logarithms (foot note)—Numerical example of branch pipe—Simplified method—Numerical Examples— Relative discharges through branches variously placed—Discharges determined by plotting—Caution regarding results obtained by calculation—Numerical examples.

In the preceding examples a series of horizontal pipes has been considered, the head being produced by an elevated reservoir placed at one end. The results would have been identical, however, if the head had been produced by the pipes being laid upon a slope, provided the difference of level between the two extremities remained the same, for the velocities and hydraulic grade line would remain unaltered. The pressure in the pipes would vary however, according to their distance below the hydraulic grade line, the pressure being measured at any given point in the pipe line, by the vertical distance between such point and the grade line. If the pipes were laid exactly upon the hydraulic grade line there would be no pressure at all in the pipes, and if they rose at any point above it, there would be either no flow or a diminished one, unless siphonage were resorted to.

In order to make this point very plain, we will consider the same system of pipes as that used in the last example, but laid as shown in Fig. 6. the upper extremity being fed by a constant supply, with only head enough to overcome resistance to entry, and produce initial velocity, which will be treated of further on

Calculating precisely as before, we get the same hydraulic

grade line, unbroken by the rising grade of the last 200 ft. of 6-in. pipe.

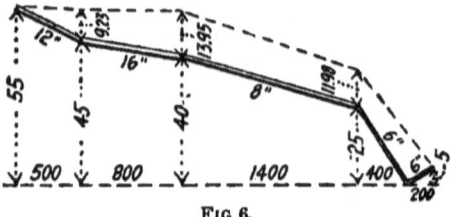

Fig 6.

It is sometimes desirable to ascertain the uniform diameter of a pipe which shall be equivalent to a series of pipes of different diameters, such as we have just been studying. This may be done by an application of formula (4), which for this purpose is written in the following form:

$$H = \frac{C Q^2}{0.616} \times \frac{L}{D^5}$$

As an example, let us calculate the diameter of a single pipe, of the same total length and fall as the series of pipes which we have just had under consideration, and capable of discharging an equal volume. We will first establish the general formula for all such problems, expressing the difference of piezometric level between the two ends of each pipe respectively, by h_1, h_2, h_3, h_4, etc., their respective lengths by l_1, l_2, l_3, l_4, etc., their respective diameters by $d_1 d_2 d_3 d_4$, etc., and their respective coefficients by $c_1 c_2 c_3 c_4$, etc., commencing with the lower end. We will express the total length by L, the total difference of level by H, the unknown diameter by D, and its coefficient by C.

Now, observing that the quantity discharged per second by each pipe is the same, we have the 4 equations:

$$h_1 = \frac{c_1 \ Q^2}{0.616} \times \frac{l_1}{d_1^5}$$

$$h_2 = \frac{c_2 \ Q^2}{0.616} \times \frac{l_2}{d_2^5}$$

PRACTICAL HYDRAULIC FORMULÆ.

$$h_3 = \frac{c_3\ Q^2}{0.616} \times \frac{l_3}{d_3^5}$$

$$h_4 = \frac{c_4\ Q^2}{0.616} \times \frac{l_4}{d_4^5}$$

Adding, and observing that the sum of the partial heads $h_1\ h_2\ h_3\ h_4$ equals H, we have :

$$H = \frac{Q^2}{0.616}\left(\frac{c_1\ l_1}{d_1^5} + \frac{c_2\ l_2}{d_2^5} + \frac{c_3\ l_3}{d_3^5} + \frac{c_4\ l_4}{d_4^5}\right)$$

but we have also the equation

$$H = \frac{C\ Q^2}{0.616} \times \frac{L}{D^5}$$

whence, suppressing the common factor :

$$\frac{C\ L}{D^5} = \frac{c_1\ l_1}{d_1^5} + \frac{c_2\ l_2}{d_2^5} + \frac{c_3\ l_3}{d_3^5} + \frac{c_4\ l_4}{d_4^5} \qquad (5$$

The above is the general formula.

Substituting the special values of our example :

$$\frac{3300}{D^5} \times C = \frac{0.33}{1^5} + \frac{0.52}{(\tfrac{3}{4})^5} + \frac{0.966}{(\tfrac{1}{2})^5} + \frac{0.432}{(\tfrac{1}{3})^5}$$

Giving a preliminary approximate value to C of 0.00066, we have

$$\frac{2.178}{D^5} = 0.33 + 0.123 + 7.335 + 13.821$$
$$D^5 = 0.1007$$
$$D = 0.63$$

This value of D indicates a practical diameter of 8 ins.

In order to check this value, we may write (4) under the form :

$$Q = \sqrt{\frac{D^5 \times H \times 0.616}{L \times C}}$$

Substituting given values :

$$Q = \sqrt{\frac{0.1007 \times 50 \times 0.616}{2.178}}$$

$$Q = 1.193 \text{ cu. ft. per second,}$$

thus proving the correctness of the work.

These calculations can be abridged, and, in many cases, sufficient accuracy secured by adopting a mean common value for C. If we do so in the present case, C becomes a common factor, and disappears from the calculation, (5) becoming

$$\frac{L}{D_*} = \frac{l_1}{d_1^{\ 5}} + \frac{l_2}{d_2^{\ 5}} + \frac{l_3}{d_3^{\ 5}}, \text{ etc.} \qquad (5)\ bis$$

If this equation be worked out for the above given values, we have:

$$D = 0.64$$

or 8 ins. as before.

It will be observed that this process might have been used with advantage in the previous example, by ascertaining the discharge of an equivalent pipe, and then calculating the heads necessary to produce this discharge through the different pipes.

In calculating fifth powers and roots, a table of logarithms is almost indispensable. If none is at hand a table of squares and cubes is of some use, remembering that a number can be raised to the fifth power by multiplying together its square and cube. Fifth roots, in the absence of logarithms, can only be extracted by "trial and error," using the above rule for fifth powers.[*]

Example 4th. A horizontal pipe (Fig. 7), 48 ins. in diameter and 2,000 ft. long, issues from a reservoir in which the surface of the water is maintained at a constant height of 50 ft. above the center of the pipe. Midway, this pipe is tapped by a branch pipe 24 ins. in diameter and 500 ft. long, with a rising grade of 4 ft. in 500. What is the piezometric head h at the junction, and what the discharge from each pipe?[†]

It is evident that the 48-in. pipe above the junction must, with the head $50-h$, discharge as much water per second as the

---

[*] All hydraulic calculations are greatly facilitated by the use of logarithms; and those engaged in making such calculations should not fail to familiarize themselves with the use of these powerful auxiliaries to arithmetical work.

[†] With these lengths and diameters, the above system does not properly come under the classification of "long pipes." As the present object is only to exemplify methods of calculation, the example is equally good.

combined discharge of the 48-in. pipe below the branch with
the head h, and the 24-in. pipe with the head $h-4$. From (3),

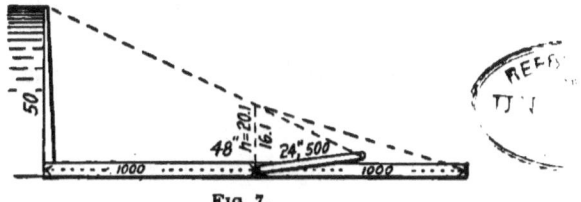

Fig. 7.

which in this case will perhaps be the most convenient equation
for quantity, though that derived from (4) is frequently useful,
we have :

$$Q = 12.56 \sqrt{\frac{4(50-h)}{1000 \times 0.00062}}$$

$$q = 12.56 \sqrt{\frac{4h}{1000 \times 0.00062}}$$

$$q' = 3.14 \sqrt{\frac{2(h-4)}{500 \times 0.00064}}$$

which, put in equation, give :

$$12.56 \sqrt{\frac{4(50-h)}{1000 \times 0.00062}} =$$

$$12.56 \sqrt{\frac{4h}{1000 \times 0.00062}} + 3.14 \sqrt{\frac{2(h-4)}{500 \times 0.00064}}$$

The coefficients 0.00062 and 0.00064 are so nearly equal that
we may, in the following calculations, discard them as common
factors. Dividing by 3.14 and striking out also the common
factors $\frac{4}{1000}$ and $\frac{2}{500}$, we have simply :

$$4\sqrt{50-h} = 4\sqrt{h} + \sqrt{h-4}$$

Squaring $\qquad 800 - 16h = 16h + h - 4 + 8\sqrt{h^2-4h}$

which gives : $\qquad 33h = 804 - 8\sqrt{h^2-4h}$

Neglecting, for a first approximate value of h the quantities
affected by the radical :

$$33h = 804$$

Neglecting decimals:
$$h = 21.$$
Substituting this value for h under the radical:
$$33h = 804 - 8\sqrt{576 - 96}$$
which gives, always neglecting decimals, a second approximate value:
$$h = 19.$$
A third and fourth approximation give respectively $h = 20.3$ and $h = 20$.

We will take 20.1 as very near the true value.*

Substituting 20.1 in place of h in the equations giving the quantities discharged, we have:
$$Q = 12.56\sqrt{\frac{4 \times 29.9}{0.62}} = 174.45$$
$$q = 12.56\sqrt{\frac{4 \times 20.1}{0.62}} = 143.05$$
$$q' = 3.14\sqrt{\frac{2 \times 16.1}{0.32}} = 31.50$$

We have thus:
$$Q = q + q'.$$

The above method gives directly the true value of h; but it involves tedious figuring, even in our example, which happens to admit of many simplifications owing to the number of common factors. It will be easier, and often shorter, to obtain the value of h by first *assuming* one which we judge likely to be near the truth, calculating what discharge it would give from the two branches, and then calculating the head necessary to discharge the same quantity from the single pipe above the branch. Then, comparing the total height thus obtained with the known height of the water in the reservoir, we can deduce the true value of h by a proportion.

Let us apply this method to the above example. We know

---

* The value of h may be obtained directly by using the usual formula for adfected quadratics; but with the aid of a table of squares and square roots, the above approximate method will generally be the easier and quicker one.

PRACTICAL HYDRAULIC FORMULÆ. 31

at once that h must be less than 25, because that would be its value if the 24-in. branch were closed. Supposing we judged that 22 ft. would be about correct. We then have to solve the two equations:

$$q = 12.56 \sqrt{\frac{4 \times 22}{0.62}} = 140.60$$

$$q' = 3.14 \sqrt{\frac{2 \times 18}{0.32}} = 33.30$$

also, for the equal discharge through the 48-in. pipe above the branch, squaring (3), we have:

$$h = \frac{(182.90)^2 \times 0.62}{(12.56)^2 \times 4} = 32.87$$

This height, added to 22, the assumed value of h, gives a total height of 54.87 ft. as against 50 ft., the actual total height. By proportion we have:

$$\frac{h}{22} = \frac{50}{54.87}$$

This value of h agrees with that already found.

If the 24-in. branch were closed we should have for the discharge:

$$Q = 12.56 \sqrt{\frac{4 \times 50}{1.24}} = 159.51$$

When the 24-in. branch was open we had a total discharge of 174.73 cu. ft. per second. There is an increase, therefore, of about 9½ per cent. by opening the branch.

Let us now see what the discharge would be if the branch were placed only 500 ft. from the reservoir, instead of 1,000 ft., all the other conditions remaining the same.

We will assume $h = 33$ ft. and solve the two equations

$$q = 12.56 \sqrt{\frac{4 \times 33}{1500 \times 0.00062}} = 149.5$$

$$q' = 3.14 \sqrt{\frac{2 \times 29}{0.32}} = 42.3$$

also

$$h' = \frac{(191.8)^2 \times 0.31}{(12.56)^2 \times 4} = 18.07$$

giving a total height of 51.07 as against 50. Reducing:

$$\frac{h}{33} = \frac{50}{51.07}$$

$$h = 32.3$$

Using this value, instead of the assumed one, we have:

$$12.56 \sqrt{\frac{4 \times 17.7}{0.31}} = 12.56 \sqrt{\frac{4 \times 32.3}{0.93}} + 3.14 \sqrt{\frac{2 \times 23.3}{0.32}}$$

$$189.83 = 148.03 + 41.76$$

very nearly.

As compared with the discharge when the 24 in. branch is closed this shows a gain of 19 per cent., just double the gain when the branch was located at the center of the pipe.

Supposing now that the branch were placed 1,500 ft. from the reservoir. Assuming 10 ft. as a probable value of h we have:

$$q = 12.56 \sqrt{\frac{4 \times 10}{500 \times 0.00062}} = 142.46$$

$$q' = 3.14 \sqrt{\frac{2 \times 6}{0.32}} = 19.23$$

also:

$$h' = \frac{(161.7)^2 \times 0.93}{(12.56)^2 \times 4} = 38.53$$

By proportion

$$\frac{h}{10} = \frac{50}{48.53}$$

$$h = 10.30$$

Using this value instead of the assumed one:

$$12.56 \sqrt{\frac{4 \times 39.7}{0.93}} = 12.56 \sqrt{\frac{4 \times 10.3}{0.32}} + 3.14 \sqrt{\frac{2 \times 6.3}{0.32}}$$

$$164.13 = 144.57 + 19.68$$

very nearly.

As compared with the discharge when the 24-in. branch is closed, this shows a gain of not quite 3 per cent., which is in marked contrast to the gain when the branch was only 500 ft.

from the reservoir, being less than one-sixth of the gain, in that case.

It will be interesting to study a little more in detail the question of relative discharges. We have seen that when there is no branch open on the 48-in. pipe, its discharge is 159.51 cu. ft. per second. The 24-in. branches, wherever placed, increase the total discharge, but diminish that in the 48-in. pipe, below the branch. By comparing the above quantities, it will be perceived that the flow from the 48-in. pipe is diminished approximately by that proportion of the quantity flowing through the 24-in. branch which is represented by its proportionate distance from the reservoir. Thus, when the branch is 1,500 ft., or three-quarters of the length of the 48-in. pipe, from the reservoir, as in the last case, its discharge is 19.62 cu. ft. per second. Three-quarters of this quantity is 14.715, which, subtracted from 159 51, leaves 144.795, or very nearly that of the 48-in. pipe below the branch, as determined by calculation.

In the same way half of the discharge, when the branch is situated half way from the reservoir, subtracted from 159.51, gives also very nearly the amount discharged below the branch. When the branch is 500 ft., or one-quarter of the total distance, from the reservoir, one-quarter of its discharge taken from 159.51 gives very closely the discharge as calculated for the 48 in. pipe below the branch.

Let us now take an extreme position for the branch, and suppose it placed close to the reservoir, so that there is practically no portion of the 48-in. pipe between it and the reservoir. There will, therefore, be no part of the flow from the branch subtracted from that of the main pipe, and the two will each discharge the same quantity as if the other were not there. That is, the 48-in. pipe will discharge 159.51, and the 24-in. 53.24 cu. ft. per second.

If we should take another extreme position for the branch, and suppose it placed at the end of the 48-in. pipe, it is obvious that, with its assumed rising grade of 4 ft. in 500, it would discharge no water at all. A position could be found by trial where it would just cease to discharge water, but for the object of the present investigation this is not necessary.

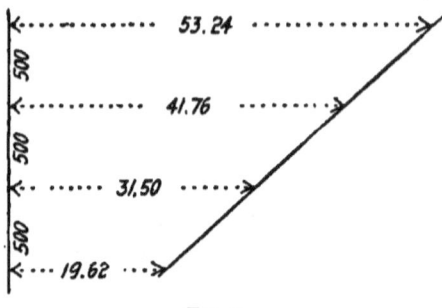

Fig. 8.

If the above results are plotted, as in Fig. 8, a very instructive diagram is obtained. The successive 500 ft. lengths being laid off as abscissæ, and the discharges measured upon the corresponding ordinates, it will be seen that their extremities all lie nearly in the same straight line. If, therefore, the discharges for any two positions of the branch be calculated, and the straight line drawn passing through their extremities, the discharge for any other position of the branch can be obtained by erecting an ordinate at the given point to the straight line, and the flow through the main also obtained by subtracting the proper portion of that of the branch.

In practice, when making calculations similar to those under consideration, one error must be carefully guarded against, namely, the supposing that the actual results will be exactly as calculated. The chief value of these calculations lies in the fact that they furnish pretty trustworthy relative results, that is, they establish fairly well in practice the fact that if a certain pipe delivers a certain volume of water in a certain position, it will deliver a certain greater or less amount in another. The actual amounts, in either case, cannot be surely determined, as they depend upon ~ many varying circumstances about which, even when aware of the existence, we have no exact data.

Let us next suppose a system in which the 48-in. pipe is tapped every 500 ft. by a 24 in. pipe, 500 ft. long, laid as before with a grade f 4 ft. in 500.

Assuming a height of 9 ft. for the piezometric column h nearest the he end of the pipe we have:

$$2.56\sqrt{\frac{4\times 9}{0.31}} + 3.14\sqrt{\frac{2\times 5}{0.32}} = 12.56\sqrt{\frac{4(h-9)}{0.31}}$$

Since the denominators under the radicals are so nearly equal we may cancel them, and making other simplifications, write:

$$\sqrt{9 + \frac{1}{8}\sqrt{10}} = \sqrt{h' - 9}$$

Whence: $h' = 20.52$

Again: $$\sqrt{11.52 + \frac{1}{8}\sqrt{33.04}} = \sqrt{h'' - 20.52}$$

$h'' = 37.43$

Also $\sqrt{16.91} + \frac{1}{8}\sqrt{66.86} = \sqrt{h''' - 37.43}$

$h''' = 63.79$

By proportion we have: $\dfrac{h}{9} = \dfrac{50}{63.07}$

$h = 7.05$

As the value of $h''' = 63.79$ differs considerably from the true value = 50, and as the above proportion is not exactly absolute, particularly in a somewhat complex system like the present, it is probable that the value just obtained for h is not a sufficiently close approximation to answer our purpose. We will therefore make a second calculation, using 7 as a second approximate value for h.

Carrying the calculation through precisely as above, we obtain the following values:

$h = 7.32$
$h' = 16.12$
$h'' = 29.60$
$h''' = 50.00$

Calculating the various discharges under these piezometric heads, calling those through the different sections of 48-in. pipe, commencing at the lower end, Q, Q', Q'', Q''', and those through the corresponding 24-in. branches, q, q', q'', we have:

$$\begin{aligned}
Q &= 122.05 \\
q &= 14.30 \\
\hline
Q + q &= 136.35 \\
Q' &= 136.10 \\
q' &= 27.62 \\
\hline
Q' + q' &= 163.72 \\
Q'' &= 163.75 \\
q'' &= 39.72 \\
\hline
Q'' + q'' &= 203.47 \\
Q''' &= 203.75
\end{aligned}$$

These results show a very close agreement.

It is worthy of note that the total discharge in this case is not greatly increased over that obtained with a single branch situated 500 feet from the reservoir. In general it will be found, as in these two cases, that when a main is tapped at a certain point by a single branch, the total discharge is comparatively but slightly increased by the introduction of a series of similar branches placed below the first junction. The position of the first branch, however, has, as the above examples show, a very great influence both on the volume of discharge and the form of the hydraulic grade line. This latter feature merits careful attention.

It will be interesting to study the effect upon the flow through such a system as we have been just considering, when the conditions are somewhat changed. For instance, in the last example let us suppose that the three branch pipes, instead of having each an equal rising grade of 4 feet in their length of 500 feet, have rising grades respectively of 4 feet, 12 feet and 24 feet in 500, commencing at the lower branch, all other conditions remaining the same.

Assuming, as before, an approximate value for h of 9 feet, we get, as before

$$h' = 20.52$$

Our next equation will be :

$$\sqrt{11.5 + \frac{1}{8}} \sqrt{17.04} = \sqrt{h'' - 20.52}$$

$$h'' = 35.81$$

Again :

$$\sqrt{15.20 + \frac{1}{8}} \sqrt{23.62} = \sqrt{h''' - 35.81}$$

$$h''' = 56.24$$

This value is sufficiently near the given one of 50, to warrant our using it to obtain pretty close approximate values, by proportion, as follows :

$$h = 8.00$$
$$h' = 18.24$$
$$h'' = 31.83$$
$$h''' = 50.00$$

Whence we obtain the following discharges

$$Q = 12.56 \sqrt{\frac{32}{0.31}} = 127.6$$

$$q = 3.14 \sqrt{\frac{8}{0.32}} = 15.7$$

$$Q + q = 143.3$$

$$Q' = 12.56 \sqrt{\frac{40.96}{0.31}} = 141.4$$

$$q' = 3.14 \sqrt{\frac{12.48}{0.32}} = 19.6$$

$$Q' + q' = 164.0$$

$$Q'' = 12.56 \sqrt{\frac{54.36}{0.31}} = 166.3$$

$$q'' = 3.14 \sqrt{\frac{15.66}{0.32}} = 22.0$$

$$Q'' + q'' = 188.3$$

$$Q''' = 12.56 \sqrt{\frac{72.68}{0.31}} + 192.3$$

This shows a pretty fair agreement between the volumes discharged, the discrepancies being due to the fact that our assumed value of h was not sufficiently close for a fine calculation. The figures are near enough, however, to serve the purpose of showing to how small an extent, comparatively, the results are changed by the very considerable changes made in the inclination of the branch pipes. Later on we shall have occasion to notice more fully the small relative changes made in the volumes discharged through given pipes by changes of grade: for the present we will only call attention to the slight variations produced in the hydraulic grade line, as determined by the piezometric heads.

CHAPTER III.

Numerical example of a system of pipes for the supply of a town—Establishment of additional formulæ for facilitating such calculations—Determinations of diameters—Pumping and reservoirs—Caution regarding calculated results—Useful approximate formulæ—Table of 5th powers—Preponderating influence of diameter over grade illustrated by example—Maximum velocities. (Note.)

As a further study of a system of pipes to deliver water, let us suppose a town divided by intersecting streets into blocks 1,000 ft. sq., as shown in Fig. 9. We will suppose that the proposed water supply requires a total volume of 3 cu. ft. per second, equal to say 800,000 U. S. galls. in 10 hours.

The water is to be introduced by a central main $A\ B\ C$, and delivered east and west by the side mains $D\ D'$, $E\ E'$, $F\ F'$, $G\ G'$, $H\ H'$. At the extremities of these mains, the water is to be delivered at the elevations above datum indicated by the figures placed in brackets. The side mains $D\ D'$ and $E\ E'$ are to deliver each, east and west, $\frac{1}{4}$ cu. ft. per second, which quantity we will suppose is to be carried through the whole length of the pipe and delivered at its extremity at the maximum elevation, without regard to the quantities drawn off *en route* by the service pipes and smaller north and south mains, nor those drawn off by the lower taps. This will secure a good delivery of water in case of fires. The total delivery of the above two side mains will therefore be 1 cu. ft. per second. The remaining three side mains, $F\ F'$, $G\ G'$, and $H\ H'$, are to deliver, similarly, $\frac{1}{3}$ cu. ft. per second at each extremity, making 2 cu. ft. for the three.

These being the data, we will suppose the problem to be the

determining of the respective diameters of the pipes, and the height to which the water must be raised in a supply reservoir or standpipe, situated somewhere to the north of the town.

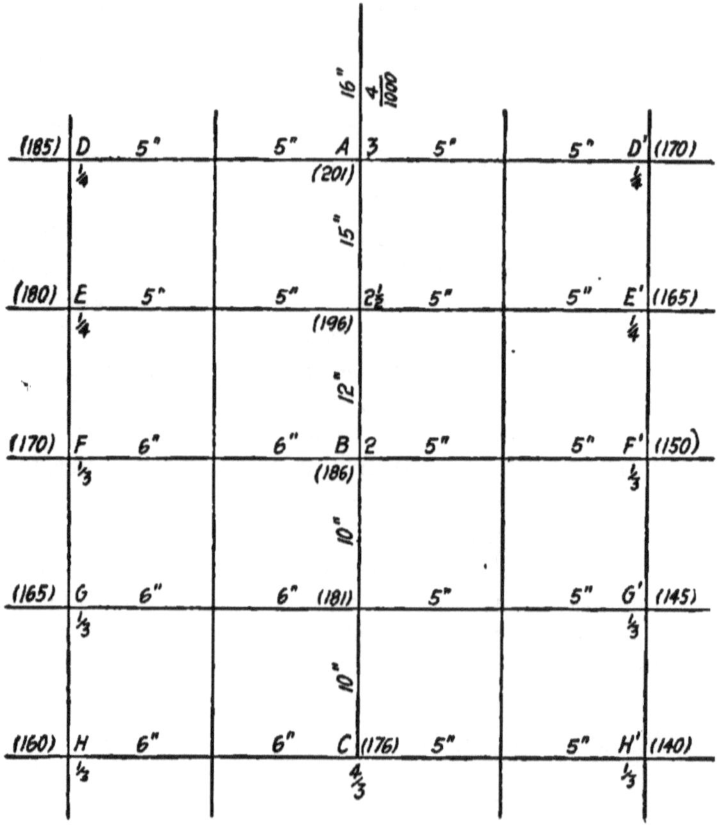

```
                    16" 4/1000

(185) D    5"       5"   A  3   5"         5"    D' (170)
      1/4              (201)                     1/4

                         15"

(180) E    5"       5"      2½   5"         5"   E' (165)
      1/4              (196)                     1/4

                         12"

(170) F    6"       6"   B  2   5"         5"    F' (150)
      1/3              (186)                     1/3

                         10"

(165) G    6"       6"  (181)   5"         5"    G' (145)
      1/3                                        1/3

                         10"

(160) H    6"       6"   C (176)  5"        5"   H' (140)
      1/3                 4/3                    1/3
```

FIG. 9a.

The problem thus stated is indeterminate and admits of an indefinite number of solutions, for we may either use large pipes and low elevations, or small pipes and high elevations. Practically, however, there are limitations to this; for in the first place we shall naturally be restricted as to the height to which it would be possible or advisable to raise the water, and secondly, experience shows that we should confine ourselves within certain limits as regards the velocity of the water in the pipes.

Generally speaking, these velocities should not exceed such as would be produced by a fall of from 4 to 8 ft. per thousand, according to the size of the pipe; the greater fall belonging to the smaller diameter. (*See note at end of chapter.*)

Before commencing the calculations, it will be well to establish certain additional formulæ, derived from (4), which are frequently of considerable utility.

When the length and diameter are constant:

$$\frac{Q'^2}{Q^2} = \frac{H'}{H}$$

When the head and diameter are constant:

$$\frac{Q'^2}{Q^2} = \frac{L}{L'}$$

When the head and length are constant:

$$\frac{Q'^2}{Q^2} = \frac{D'^5 C}{D^5 C'}$$

$$\frac{D'^5}{D^5} = \frac{Q'^2 C'}{Q^2 C}$$

When the head and discharge are constant:

$$\frac{D'^5}{D^5} = \frac{L' C'}{L C}$$

When the length and discharge are constant:

$$\frac{D'^5}{D^5} = \frac{H C'}{H' C}$$

These relations indicate that, other things being equal, the squares of the discharges vary directly as the heads and the fifth powers of the diameters, and inversely as the lengths; and that, other things being equal, the fifth powers of the diameters vary directly as the squares of the discharges and the lengths, and inversely as the heads.

As these relations are generally used for approximations, the coefficients may be dropped, and the equations written in this form:

$$Q' = \sqrt{\frac{Q^2 \times H'}{H}} \quad (6)$$

$$Q' = \sqrt{\frac{Q^2 \times L}{L'}} \quad (7)$$

$$Q' = \sqrt{\frac{Q^2 \times D'^5}{D^5}} \quad (8)$$

$$D' = \sqrt[5]{\frac{D^5 \times Q'^2}{Q^2}} \quad (9)$$

$$D' = \sqrt[5]{\frac{D^5 \times L'}{L}} \quad (10)$$

$$D' = \sqrt[5]{\frac{D^5 \times H}{H'}} \quad (11)$$

Other combinations can be made from these relations. Thus:

$$D' = \sqrt[5]{\frac{D^5 \times H \times Q'^2}{H' \times Q^2}} \quad (12)$$

Commencing now with the west side of the main $H H'$, we have $\frac{1}{3}$ cu. ft. to be delivered at an elevation of (160) above datum. As the pipe will be a comparatively small one, we will assume a grade of $\frac{8}{1000}$, which will give a rise of 16 ft. between the extremity and the main junction, and requires an elevation of piezometric head, at this junction, of (176), as shown in the figure.

To obtain the proper diameter of pipe for this grade and discharge, we have, using (4), and assuming $C = 0.00076$ as a probable value;

$$D^5 = \sqrt{\frac{(\tfrac{1}{3})^2 \times 1000 \times 0.00076}{8 \times 0.61}}$$

whence $D^5 = 0.017304$
and $D = 0.444$.

PRACTICAL HYDRAULIC FORMULÆ.   43

• Or, for the next highest even inch :

$D = 6$ inches.

As regards the diameter of the pipe on the east side, since the length and discharge are the same as for the west side, and only the heads vary, being respectively 16 and 36 ft., it can be obtained by means of (11).

Thus :

$$D' = \sqrt[5]{\frac{0.017304 \times 16}{36}}$$

$D' = 0.3777$

or, for next highest even inch :

$D' = 5$ inches.

The above head of 18 ft. per thousand produces a velocity of flow in a 5 in. pipe of a little over 3 ft. per second, which is somewhat greater than it should be. If the limit of velocity is overstepped to any considerable degree in a system of pipes such as we are considering, it would be best to use a larger pipe and check its flow down to the desired delivery by means of a gate or stop cock placed near its upper end, the effect of which will be to diminish the head. In the present instance the excess of velocity is probably not sufficient to render this precaution necessary.

The elevations are such that the above diameters of 6 and 5 ins. are also proper for the side mains $G\ G'$, $F\ F'$.

It is now necessary to calculate the diameter of the central main from B to C. This main might be divided into two parts, that between $F\ F'$ and $G\ G'$ and that between $G\ G'$ and $H\ H'$, but we will calculate it upon the supposition of a uniform diameter, capable of delivering the entire volume of $\frac{3}{4}$ cu. ft. per second as far as $H\ H'$.

Assuming a probable value of $C = 0.00066$, we have from (4):

$$D^5 = \frac{\frac{16}{9} \times 1.32}{6.1}$$

whence:
$$D^5 = 0.3817$$
and:
$$D = 0.826 = 10 \text{ ins.}$$

Taking now the mains $E\,E'$ and $D\,D'$, and beginning on the west side, assuming as before a grade of 8 ft. per 1,000, we find the length and head equal to those of $F\,F''$ etc., the only difference being the quantity it is desired to deliver, which is now ¼ cu. ft. as against ½ in $F\,F'$. The relation (9) is therefore applicable, and we have:

$$D = \sqrt[5]{0.017304 \times \frac{1}{16} \over \frac{1}{9}}$$

whence:
$$D^5 = 0.0097335$$
and
$$D' = 0.396$$
or, say,
$$D' = 5 \text{ ins.}$$

The mains on the east side are determined as before:

$$D' = \sqrt[5]{0.0097335 \times \frac{16}{31}}$$

$$D' = 0.346$$

This is not quite 4¼ ins., but to insure the desired delivery, it will be best to take the next highest even inch, and call it 5 ins.

As regards the central main from A to B, we find two grades, the upper one $\frac{6}{1000}$ and the lower $\frac{10}{1000}$. The lower section must deliver, under a grade of $\frac{10}{1000}$, all the water required for $F'F''$, $G\,G'$, and $H\,H'$, aggregating 2 cu. ft. per second. Using (4), and taking 0.00066 as a probable value of C, we have:

$$D^5 = \frac{4 \times 0.66}{6.1}$$

whence :
$$D^5 = 0.4328$$
and :
$$D = 0.846$$

This is very nearly 10¼ ins., and a 10 in. pipe would answer, though 12 ins. would be better.

The upper section must deliver 2.5 cu. ft. per second, under a grade of $\frac{5}{1000}$. Taking the same probable value of C, we have :

$$D^5 = \frac{6.25 \times 0.66}{3.05}$$

whence :
$$D = 1.237$$

which we can take as either 15 or 16 ins.

This diameter might have been obtained from that of the lower section, by means of (12). Thus :

$$D'^{\,5} = 0.4328 \times \frac{10}{5} \times \frac{6.25}{4}$$
$$D' = 1.287$$

This last formula might have been used throughout, but (4) is about as short and convenient ; frequently more so.

The diameters being thus determined, the quantities should be verified by (3). They will be found somewhat in excess of those proposed, owing to the general increase of the diameters.

As regards the height to which the water must be raised, the data show that 3 cu. ft. per second must be raised to a sufficient height to reach DD' at an elevation of (201) above datum. If we adopt a grade of $\frac{4}{1000}$, the proper diameter of the pipe would be :

$$D^5 = \frac{9 \times 0.65}{2.44}$$
$$D = 1.32$$

or,
$$D = 16 \text{ ins.}$$

If, instead of pumping, the water were collected in a reservoir by damming up the natural flow of some stream, and the dam were of necessity situated at an elevation so great that a dangerous pressure is apprehended, it would be necessary to first receive the water into a distributing reservoir situated at a lower level, or else, as a less advantageous expedient, to reduce the pressure by gates, properly located for the purpose.

It should be well understood that all the above assumed data, particularly such as relate to heads, are subjected to considerable variation in actual practice. All the calculations have been based, of necessity, upon the hypothesis that the exact allotted volume per second is being simultaneously drawn from the whole system. This would rarely be the case; for at any given second, the draught would be liable to fluctuate greatly from the average. Indeed, these calculations should only be regarded as fixing, with some degree of approximation, the proper relative discharges and pressures at the different points supplied.

The remaining north and south pipes should be calculated in the same way. Thus, those below $F\,F'$ on the west side discharge 1·6 cubic ft. with a grade of $\frac{8}{1000}$. This would require a 4 in. pipe. The draught from these would somewhat lower the piezometric heads at their junctions with the side mains. In a fine calculation, these reductions should be worked out, as was done in the previous example of branch pipes; in general, however, and in cases where the whole supply is supposed to be carried through to the extremity of the mains, and delivered at the highest elevation, as was done in the present instance, and where a liberal interpretation has been given to the calculation of diameters, this is not indispensable. At the same time, it should be a guiding principle of water-works engineering that a few hours spent in the office, in what may sometimes be considered an over-refinement of calculation, is by no means a waste of time, and frequently enables one to make advantageous and economical modifications in a project of distribution.

It may here be noted that (12) admits of being put into a very convenient form for rapid approximations. To do this, we have only to calculate the discharge of a pipe 1 ft. in diameter, with a fall of 1 ft. per thousand, and to refer all other discharges with the fall per thousand feet to it, in order to obtain the corresponding diameter. The quantity discharged by the above pipe is 0.961 cu. ft. per second, and the square of the same is 0.924. Equation (12) may then be written:

$$D = \sqrt[5]{\frac{Q^2}{H}} \times 1.08$$

or very nearly:

$$D = \sqrt[5]{\frac{Q^2}{H}} \tag{13}$$

we have also very nearly:

$$Q = \sqrt{D^5 \times H} \tag{14}$$

which may be more conveniently expressed thus:

$$Q = D^2 \sqrt{D \times H} \tag{14 bis}$$

We have, also,

$$V = \sqrt{D \times H \times 1.6} \tag{14 ter.}$$

in which V = velocity in feet per second.

These last formulæ, it will be perceived, are based on the fact that, given a certain probable degree of roughness, a pipe 1 ft. in diameter, with a fall of 1 ft. in a thousand, will deliver 1 cu. ft. of water per second. If we desire to apply them to smooth, clean pipes, we have only to *halve the coefficient* for a 12-in. pipe, which will be equivalent to writing the above formulæ thus:

$$D = \sqrt[5]{\frac{Q^2}{2H}} \tag{15}$$

$$Q = \sqrt{D^5 \times 2H} \tag{16}$$

These formulæ will be found of very great utility in arriving quickly at approximate results. They can be advantageously used in sketching out a network of pipes such as we have just been considering. To facilitate their use the following table of fifth powers has been calculated. This table indicates, by inspection, the diameters in inches corresponding to the fifth roots of the right-hand side of the equations, expressed in feet.

Diameters in inches.	Fifth Powers in feet.	Diameters in inches.	Fifth Powers in feet.
3	0.000977	22	20.72
4	0.004115	24	32.00
5	0.01256	26	47.75
6	0.03125	28	69.17
8	0.1317	30	97.66
10	0.4019	32	134.9
12	1.000	34	182.6
14	2.1615	36	243.0
16	4.214	40	411.5
18	7.594	42	525.2
20	12.86	48	1,024.0

All the diameters which have been already calculated can be obtained very nearly by the use of (13). Relations (13) and (14) might also have been used in some of the previous examples.

Formulæ (13) and (14) serve to show the comparatively small influence of *grade* as affecting the volumes discharged, which point has been already alluded to, and the preponderating influence of *diameter*. Thus, we see by the above formulæ, that for a diameter of 1 ft. and a fall of $\frac{1}{1000}$, the volume of discharge is 1 cu. ft. If we wish to double this discharge by increasing the fall, we must adopt a grade of $\frac{4}{1000}$, i. e., we must quadruple the fall. If, on the other hand, we wish to produce the same result by increasing the diameter without changing the grade, we need only adopt a diameter of 1.32 ft. and even a little less, on account of the decrease in the coefficient. That is to say, to double the discharge, we must increase the fall 300 per cent., or the diameter 32 per cent.

NOTE.— In completion of what has been already said in this chapter (page 41), regarding the limit of velocities for pipes of different diameters, the following table (founded upon that given by Mr. Fanning) indicates pretty closely the maximum velocities which it is generally advisable to produce :

Diameter in inches.........	6	12	18	24	30	36	42	48
Velocity in ft. per sec......	2.5	3.5	4.5	5.5	6.5	7.5	8.5	9.5

CHAPTER IV.

Use of formula (14) *illustrated by numerical example of compound system combined with branches—Comparison of results—Rough and smooth pipes—Pipes communicating with three reservoirs—Numerical examples under varying conditions—Loss of head from other causes than friction - Velocity, entrance and exit heads—Numerical examples and general formulæ—Downward discharge through a vertical pipe—Other minor losses of head—Abrupt changes of diameter—Partially opened valve—Branches and bends—Centrifugal force—Small importance of all losses of head except frictional in the case of long pipes—All such covered by "even inches" in the diameter.*

As an illustration of the use of (14) we will calculate by its aid the discharge from a reservoir, tapped at a depth of 50 ft. by a horizontal compound system consisting successively of 2,000 ft. of 12-in. pipe, 2,000 ft. of 24-in. pipe and 2,000 ft. of 12-in. Each of these three lengths of pipe is tapped midway by a 6-in. pipe, laid horizontally, the one nearest the reservoir having a length of 3,000 ft.; the next 1,000 ft., and the last 500 ft. (See Fig. 9, *bis*.) All the pipes being open, it is desired to find the piezometric heads h, h', h'', h''', h'''', at each branch and change of diameter, and the volumes discharged by each branch and section of main pipe.

Beginning at the lower end and assuming 6 ft. as an approximate value of h, we have from (14), H always representing the fall per 1,000:

$$\sqrt{6} + \sqrt{\frac{12}{32}} = \sqrt{h' - 6}$$
$$h' = 15.36$$
$$\sqrt{9.36} = \sqrt{32(h'' - 15.36)}$$
$$h'' = 15.65$$
$$\sqrt{9.36} + \sqrt{\frac{15.65}{32}} = \sqrt{32(h''' - 15.65)}$$

$$h''' = 16.09$$
$$\sqrt{14.08} = \sqrt{h'''' - 16.09}$$
$$h'''' = 30.17$$
$$\sqrt{14.08} + \sqrt{\frac{10.06}{32}} = \sqrt{h''''' - 30.17}$$
$$h''''' = 48.82$$

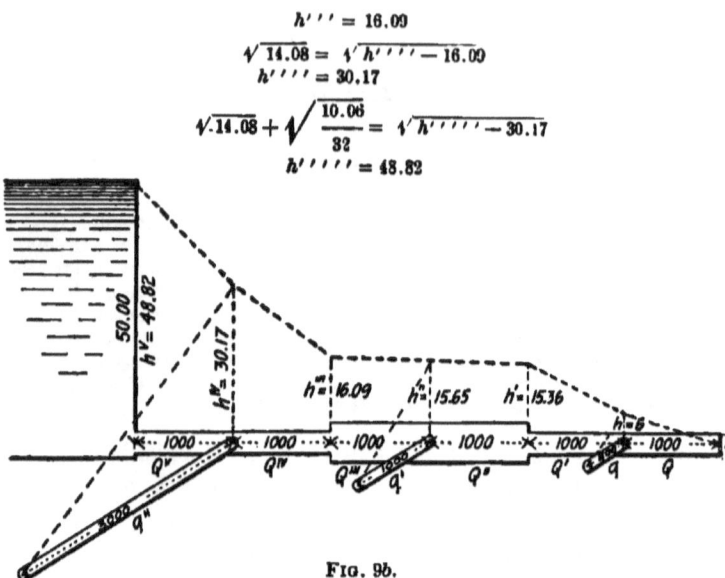

Fig. 9b.

Comparing this value with the given height 50, we may increase all the preceding values of h, h', etc., in the proportion of $\frac{50}{48.82}$. But in practice we would not wish to reckon on the total head, and it would be preferable therefore to let the values stand as they are.

We will now calculate the quantities, calling those discharged from the successive sections of main pipe, beginning at the lower end, Q, Q', Q'', Q''', Q'''', and Q''''', and those discharged by the branches, beginning also at the lower end, q, q', q'', respectively, using both (3) and (14). The results given by (14) naturally check exactly, since they depend directly upon the method used in determining h, h', etc.

	By (3)	By (14)
Q	= 2.39	2.45
q	= .56	.61
$Q + q$	= 2.95	3.06

	By (3).	By (14).
Q'	= 2.96	3.06
Q''	= 2.99	3.05
q'	= .65	.70
$Q'' + q'$	= 3.64	3.75
Q'''	= 3.68	3.75
Q''''	= 3.63	3.75
q'''	= .52	.56
$Q'''' + q''$	= 4.15	4.31
Q'''''	= 4.19	4.32

The above example was very favorable to the use of (14), because of the lengths assumed for the different pipes, but in almost all cases it will greatly reduce the volume of calculation, and frequently give sufficiently close results. Indeed, as all these calculations are merely approximations, and as we have taken our coefficients pretty high, it would no doubt often be found, could the actual discharges be measured, that the apparently less exact formula gave the more correct results.

In all the previous examples, the coefficients for rough pipes have been used. It is well to remember that, as is shown by (15) and (16), the discharge of a clean pipe of given diameter is about 41 per cent. greater than that of a rough pipe of the same diameter; also that the diameter of a clean pipe, discharging an equal volume with a rough one, will be about 88 per cent. of the latter. Between these limits of smoothness and roughness there are, of course, an indefinite number of gradations.

A very interesting investigation is that of a system of pipes communicating with two reservoirs, and discharging either freely in the air, or into a third reservoir situated at a lower elevation as shown in Fig. 10.

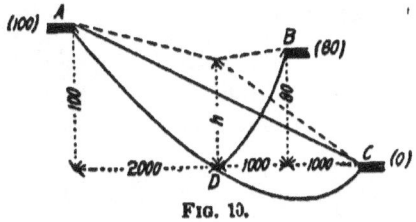

Fig. 10.

Let us suppose the water surfaces in A and B to be respectively 100 and 80 ft. above the water surface in C, and that all the pipes shown in the figure are 12 ins. in diameter. Let the total length of pipe from A to C be 4,000 ft.

If communication were shut off from B, the flow would be direct from A to C; if communication were shut off from C, it would be direct from A to B. If A were shut off, the flow would be from B to C. If all the communications were wide open, we desire to know whether the flow would be from A to B and C, or from A and B to C; and in either case, to know the piezometric head h, at the junction D, and the volumes discharged.

First, let the junction D be situated midway in the 4,000-ft. pipe joining A and C, and let the length BD be 1,000 ft. Let us for a moment revert to the supposition that B is shut off. The flow would then be from A to C, the hydraulic grade line would be a straight line joining the surfaces A and C, and under our present hypothesis, that the junction D is in the middle of AC, the piezometric head h would be 50 ft. above the surface of the lower reservoir C. But B is supposed to be 80 ft. above the same, and therefore the flow must be from A and B to C. We might at first sight suppose that the flow from B to C would be in virtue of the head $80 - 50 = 30$ ft., which is the difference of level between B and the piezometric head at the junction; but just as a branch drawing water *from* a main pipe lowers the piezometric head at the junction, so does a branch discharging *into* the main pipe raise it. It is necessary to see what the height h will be in the present case.

The quantity discharged into C is equal to the sum of the quantities passing from A and B. All areas and coefficients being equal, and all reductions made, we have:

$$\sqrt{\frac{h}{2}} = \sqrt{50 - \frac{h}{2}} + \sqrt{80 - h}$$

PRACTICAL HYDRAULIC FORMULÆ. 53

whence :
$$h = 65 + \sqrt{4000 - 90h + \frac{h^2}{2}}$$
and, by successive approximations :
$$h = 74$$
Using this value of h in (3), we obtain the different discharges as follows :
$$Q = 5.88$$
$$Q' = 3.18$$
$$Q'' = 2.37$$
This gives a very close agreement in the relation $Q = Q' + Q''$.

Suppose now that the diameter of the branch BD be reduced to 6 in., all the other conditions remaining the same. Still regarding the coefficients as equal, in order to get rapidly at an approximation, factoring the areas and simplifying, we have :
$$4\sqrt{\frac{h}{2}} = 4\sqrt{50 - \frac{h}{2}} + \sqrt{40 - \frac{h}{2}}$$
whence :
$$16.5h = 840 + 4\sqrt{8000 - 180h + h^2}$$
and, by successive approximations :
$$h = 58$$
This value of h gives the following quantities :
$$Q = 5.21$$
$$Q' = 4.43$$
$$Q'' = 1.08$$
A tolerably close check, but showing that the true value of h is a little greater than the even 58 ft. at which we have placed it.

Let us now suppose that the pipe BD is increased to a diameter of 36 in., all the other conditions remaining as before.

Then :
$$\sqrt{\frac{h}{2}} = \sqrt{50 - \frac{h}{2}} + 9\sqrt{80 - h}$$
whence :
$$h = 79.90$$
Giving :
$$Q = 6.111$$
$$Q' = 3.065$$
$$Q'' = 2.816$$

a close approximation; the true value of h lies between 79.85 and 79.90.

As h increases with the diameter of the pipe BD, it might at first seem as though, by indefinitely increasing the diameter, h might be so increased as to cause a flow from A into B. A moment's reflection, however, will show that under the assumed conditions the diameter can never be sufficiently increased to cause a flow toward B. For it has been seen that when B is shut off, the piezometric head at D is 50 ft. It is raised by opening the communication with B, and allowing water to flow into the main from B. It is evidently, therefore, an essential condition of the increase of piezometric height that the flow should be from, not to, the reservoir B.

But the effect will be different if the junction D be sufficiently advanced toward the reservoir A. Let us suppose the positions of the three reservoirs to remain the same, all the pipe diameters to be 12 ins., and the point of junction of the pipe BD to be placed at 500 ft. from A (Fig. 11). If communication with B were shut off, the piezometric height at D would be 87.5 ft. There would therefore be a flow from A to B and C when the pipe leading to B was open. But this flow would not take place under the head 87.5, for the draft toward B would lower it.

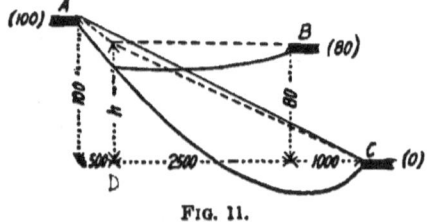

FIG. 11.

To ascertain the true value of h at the point D, we have the relation:

$$\sqrt{\frac{100-h}{500}} = \sqrt{\frac{h}{3500}} + \sqrt{\frac{h-80}{2500}}$$

simplifying:

$$\sqrt{100-h} = \sqrt{\frac{h}{7}} + \sqrt{\frac{h-80}{5}}$$

$$47 h = 4360 - 11.86 \sqrt{h^2 - 80 h}$$

whence, by successive approximations:

$$h = 82.65$$

Using this value of h we get:

$$Q = 5.695$$
$$Q' = 4.698$$
$$Q'' = 0.995$$

When B is shut off, in the above system, the discharge from A to C is 4.83 cu. ft. per second.

In all that precedes, only the resistance due to friction has been considered, and the total difference of level between the source of supply and the discharge has been taken as available for overcoming this frictional resistance. In the case of long pipes, where the velocity is comparatively low, this resistance is so greatly in excess of all the others that, in order to simplify calculations, they are neglected. This leads to no material error in cases where the pipe is over 1,000 diameters in length.

Attention, however, has been already called to the fact that there are other resistances which require a certain proportion of the total head to overcome them, leaving only the remainder available as against friction. Indeed, it is evident, if we assume all the head to be consumed by frictional resistance alone, the water in the pipe would be in exact equilibrium, and no flow could take place.

It will now be proper to show how the total loss of head, from all causes, may be calculated. And first, a word in reference to the phrase "loss of head" just employed. This term, often met with in treatises on hydraulics, may occasionally prove confusing. It is really little more than a convenient abbreviation. When we speak, for instance, of "the loss of head due to velocity," we mean the head, or fall, theoretically necessary to produce that velocity. Similarly, when we speak of "the loss of head due to resistance to entry," we mean the amount of head, or pressure, necessary to force the fluid vein into the mouth of the pipe or orifice, against the resistance of its edges. This resistance, it may be remarked in passing, as well as that due to bends, elbows and branches, shortly to be mentioned, is caused by the fact that water is not a perfect fluid, and therefore changes of direction in its flow require a certain amount of force to break or distort the form of the fluid vein as, though to a very much less degree, would be the case with a plastic body under similar circumstances. The property of water which causes these resistances is called its *viscosity*.

As applied to long pipes, the principal "loss of head," and the only one hitherto considered, is the *frictional*. The term thus applied means the height or pressure necessary to overcome the friction of the water passing with a given velocity through a pipe of given length and diameter. Thus, when we speak of the frictional loss of head per 1,000 ft. in reference to a given pipe, we mean the fall per 1,000 ft. necessary to maintain the given or desired velocity, as against friction.

We will now investigate this subject by means of the following problem: Two reservoirs (Fig. 12) containing still water and having a difference of level of 30 ft., are joined by a pipe 12 ins. in diameter and 3,000 ft. long. What is the velocity of discharge between the upper and lower reservoirs?

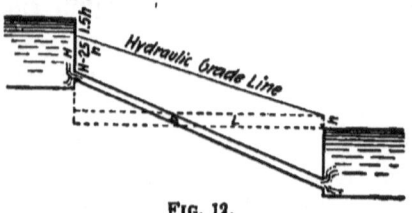

FIG. 12.

From what has been already said, it will be seen that, besides the frictional loss of head, there will be the loss of head due to

PRACTICAL HYDRAULIC FORMULÆ. 57

velocity and that due to entrance. If the pipe discharged freely in the air at its lower end, at the vertical distance of 30 ft. below the surface of the water in the upper reservoir, these three would be the only losses of head incurred, and their sum would be equal to 30 ft.; but as the discharge takes place in a reservoir, the surface of the water in which is supposed to cover the end of the pipe, to a sufficient depth to cause the discharge to take place in still water, there is the further loss of head due to the *extinction of the velocity* which is dissipated in vortices. This loss constitutes what may be called the *back pressure* of the reservoir.

In solving this problem, let us first, as heretofore, neglect all losses except frictional ones. We have then, from (1), using the above data, and the coefficient for rough pipes :

$$\frac{1}{100} = 0.00066 \, V^2$$
$$V^2 = 15.15$$
$$V = 3.89 \text{ ft. per second.}$$

The head theoretically necessary to produce this velocity is given by the formula derived from the law of falling bodies, $h = \frac{V^2}{2g}$ by substitution of the above value V. Thus :

$$h = \frac{15.15}{64.4}$$
$$h = 0.2352$$

Besides this, there is the loss of head due to entrance. We have already seen that this is always equal to about half the velocity head. We have then :

$$h + \frac{h}{2} = 0.3528$$

The loss of head from back pressure of the water in the lower reservoir, being that necessary to extinguish the velocity, must be equal to that necessary to produce the same. We have, therefore, for the total losses, outside of friction :

$$h + \frac{h}{2} + h = 0.588$$

And the head available for overcoming friction becomes

$$30 - 0.588 = 29.412$$

We must now recast our original calculation, using 29.4 ft. instead of 30 as available frictional head. Thus:

$$\frac{29.4}{3000} = 0.00066 \, V^2$$
$$V^2 = 14.8$$
$$V = 3.85$$

This is a very small reduction from the velocity already obtained. But, in order to see how our previous solution is affected by the change, we will work out new values for the subheads. Thus:

$$h = \frac{14.8}{64.4}$$
$$h = 0.23$$
$$h + \frac{h}{2} + h = 0.575$$
$$30 - 0.575 = 29.425,$$

leaving the previous value practically unchanged.

Let us now see, by means of a general formula, what is the amount of error which we commit when we ignore all resistances except friction.

Calling V the actual mean velocity, that is the actual volume discharged divided by the area of the pipe (3), we have, in the case of discharge between two reservoirs, as shown in Fig. 12, the following subheads, which together make up the total head H:

$$H = \frac{V^2}{2g} + \frac{V^2}{4g} + \frac{V^2}{2g} + \frac{L C V^2}{D}$$
$$H = \frac{5 V^2}{4g} + \frac{L C V^2}{D}$$
$$H = 0.039 \, V^2 + \frac{L C V^2}{D}$$

That is to say, by using (3), which gives

$$H = \frac{L C V^2}{D}$$

we make the error of omitting a height not quite equal to 4 per cent. of the square of the velocity.

In long pipes this is a very trifling amount.

If the pipe discharged in free air, we would have:

$$H = \frac{V^2}{2g} + \frac{V^2}{4g} + \frac{L C V^2}{D}$$

$$H = 0.0233\, V^2 + \frac{L C V^2}{D}$$

In this case we make the still smaller error of omitting $2\frac{1}{3}\%$ of V^2.

In all cases, having obtained V^2 by means of (1), we can easily judge from the nature of the problem whether it is necessary to take account of these errors. In designing a system of pipes, where the problem generally is to find the proper diameter for a certain discharge, the practice of taking the next highest even inch will almost always amply suffice to cover all omissions.

As has been already stated, in all ordinary circumstances of pipelaying, the horizontal measurement of the pipe is taken instead of its actual length. It is only in special cases that this cannot be done. The extreme limit occurs in the case of a vertical pipe discharging from the bottom of a reservoir. This constitutes a very interesting special case, for should the reservoir be of indefinitely large area, but of relatively shallow depth, the relation $\frac{H}{L}$ tends toward unity as L, and consequently H, increase.

The velocity, as determined by (1), tends therefore toward:

$$V = \sqrt{\frac{D}{C}}$$

and remains constant, no matter how greatly L may be increased. If we apply this formula to a 12-in. pipe of indefinite length, using the coefficient for rough pipes, we get,

$$V = 38.9$$

This is the maximum velocity of discharge in feet per second for a vertical 12-in. pipe under the given circumstances.*

There are several minor losses of head, besides those already considered, which are liable to occur from changes of diameter, branches, and bends or elbows. Our experimental knowledge of the effects of these features is very limited, and it is probable that much weight should not be attached to the formulæ given for their determination. A brief space will be devoted to their consideration, more with a view to make the present paper complete than for any practical value which they possess.

When water passes through a pipe of which the diameter is abruptly changed, at a certain point, to a greater or a smaller one, there is a loss of head due to the eddies formed and the sudden contraction of the fluid vein. In practice such pipes are always joined by a *reducer*, or special casting, which forms a tapering connection between the two. This greatly diminishes the agitation of the water in passing from one pipe to the other. It would seem, however, that the mere change of velocity, independent of such agitation, causes some slight modification of the profile of the hydraulic grade line; and it will be well, in any event, to give formulæ for the different cases which may occur when abrupt changes take place, as these give rise to the maximum retardation. The following formulæ are taken from Claudel's *Aide Mémoire*, ninth edition.

First.—When the change is from one pipe to another of smaller diameter, we have:

whence:
$$h = 0.49 \frac{V^2}{2g}$$
$$h = 0.00076 \, V^2$$

V being the velocity of the water in the smaller pipe. We have seen, by examples previously given, how this velocity may be obtained.

---

* The same result may be inferred from what has been said in Chapter I. about a pipe laid so that its axis coincides with the hydraulic grade line. Obviously, a vertical pipe discharging downward is a special case of such coincidence.

PRACTICAL HYDRAULIC FORMULÆ.   61

Second.—If the water (Fig. 13), in its passage from the greater to the smaller pipe, passes through an opening in a thin diaphragm,

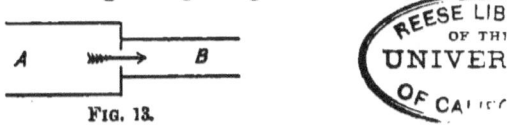

FIG. 13.

as in the case of a partially opened stop-cock, we have :

$$h = \frac{V^2}{2g} \cdot \left(\frac{S}{0.62 S'} - 1 \right)^2$$

in which V is the velocity in B, S the area of cross-section of B, and S' the area of the opening in the diaphragm.

Third.—When the flow is from one pipe to another of larger diameter :

$$h = \frac{(V - V')^2}{2g}$$

in which $V =$ velocity in small pipe, and $V' =$ velocity in larger one. When the water passes from a pipe into a reservoir, as in the case lately considered, V' becomes zero, and we have, as already established in that case :

$$h = \frac{V^2}{2g}$$

Another loss of head is that due to branches (Fig. 14). In

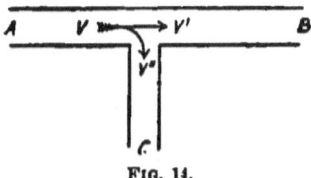

FIG. 14.

this case the water flowing from A, with a velocity V, is split at the junction, part passing on toward B, with a reduced velocity V', and part entering the branch and flowing toward C, with the velocity V''. The loss of head occasioned by perturbations of the water at the junction has not been satisfactorily investigated.

When the branch leaves the main at a right angle, this loss, as determined by a few incomplete experiments, is :

$$h = \frac{3\,V''^2}{2g}$$

V'' being the velocity in the branch. We have already seen how this velocity may be calculated.

If, as is generally the case in practice, the branch is deflected gradually instead of forming an abrupt angle of 90°, the vortices are nearly annulled, and the only loss can be from the difference of the velocities in the three pipes. Thus for B and C, respectively, we have :

$$h = \left(\frac{V - V'}{2g}\right)^2$$

$$h' = \left(\frac{V - V''}{2g}\right)^2$$

For bends, or elbows, Navier's formula for loss of head is :

$$h = \frac{V^2}{2g}\left(0.0123 + 0.0183\,k\right)\frac{A}{R}$$

in which $V =$ velocity of flow, $R =$ the radius of the bend, taken along the axis of the pipe, and $A =$ the length of the bend, also measured along the axis.

It will readily be seen how very trifling the loss of head from this cause will be in all ordinary cases.

The water passing around a bend exercises a radial thrust upon it which may sometimes be so considerable as to require bracing against. The expression for the centrifugal force F is :

$$F = \frac{M V^2}{R}$$

in which $M =$ the mass of the liquid in motion, $V =$ its velocity, and $R =$ the radius of the bend measured on its axis.

As an illustration, we will suppose a pipe 24 ins. in diameter, through which the water flows with the velocity of 8 ft. per second, around a bend of 8 ft. radius.

The mass of the liquid in motion is its weight divided by g. The centrifugal force, therefore, per running foot is :

$$F = \frac{3.14 \times 62.5}{32.2} \times \frac{8^2}{8}$$

$$F = 48.72 \text{ lbs.}$$

If the bend turns a quarter circumference, its development on the axis will be 12.57 ft., and the total thrust on the bend will be $48.72 \times 12.57 = 612.4$ lbs.

This would be liable to be intensified by sudden changes in velocity, and if the bend is not well abutted, might tend to draw the joints.

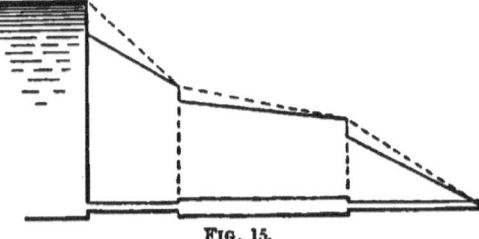

FIG. 15.

Fig. 15 shows the manner in which such losses of head as we have been just considering modify the the profile of the hydraulic grade line. The dotted line shows the grade as determined by the calculations which we have already made for a line of pipes of varying diameter. The full line, broken at the reservoir and at each change of diameter, shows the hydraulic grade as modified by losses of head due to velocity and changes of diameter. It will be understood, of course, that this is a mere random sketch, without reference to proportion.

The result of what precedes in reference to all losses of head other than friction shows that in practice, and in the case of long pipes, such losses exercise but a trifling influence. A very small increase in the diameter of the pipe over that obtained by calculation based on frictional head alone, such as would naturally be made to get even inches, will in almost all cases largely cover all losses due to velocity, entrance, branches, bends, etc.

CHAPTER V.

Notes on Pipelaying.

It will not be amiss at present to give some hints respecting *Pipelaying*. Enough has been already said to show how greatly the smoothness or roughness of the interior of a pipe affects the velocity of the flow of water through it. A line of pipes is made up of a great number of separate lengths joined together, generally by the *spigot end* of each pipe entering into the *hub* or *bell end* of the other. Each of these joints occasions more or less friction, and it is essential, not only on this account, but also and more particularly in order to make a substantial and enduring piece of work, that the pipes should be laid as evenly as possible, and the joints well fitted and calked. The alinement should be straight and the grade regular. This latter is the more important of the two, because sags and depressions in the line occasion deposits of impurities in the low points and accumulations of air in the high ones. The line should run straight and even between changes of grade and direction. Each low point, or point from which the grade rises both ways, should be provided with a special "blow-off" and stop cock, to clear it of sediment by blowing off under pressure, and each summit, or point from which the grade falls both ways, should have a special air vent, or hydrant, to discharge the accumulated air from time to time. When a new line of main is filled for the first time, or when a line is refilled after having been emptied for any cause, all the blow-offs and air cocks should be opened, and the water admitted very slowly by giving a few turns only to the admission valve. Then, as the pipe gradually fills, the blow-offs should be closed progressively as the water reaches them, and the air cocks also, beginning to close the latter, if possible, from the lower end, and only when they discharge *solid water*. Changes of horizontal direction should be joined by as easy curves as can be obtained. In sharp curves and large diameters, special curved pipe may be necessary, but, in general, curves are got in with straight pipe, using all short pieces that may be on hand, and, if necessary, cutting whole pipe, and joining the straight pieces with sleeves.

In a well-laid pipe line, all pipe, particularly all those of 20" diameter and upward, are laid on blocks. These blocks consist of pieces of wood sawed out to regular dimensions, there being two under each pipe, one just behind the hub and the other as near the spigot end as will permit of the joint being easily reached for calking, say about 2 feet. For diameters from 36" to 48", the length of these blocks may be equal to the diameter of the pipe, and about a foot wide and 6" thick. For smaller pipe they may be about two feet longer than the diameter of pipe and proportionately lighter than for the larger sizes. The pipe is held in its place on these blocks by means of wooden wedges placed side by side, on opposite sides of the pipe, and driven past each other. For 48" pipe these wedges may be about 18" long, 6" wide and 4" thick at the butt. For smaller pipe they will of course be lighter.

The instrumental alinement of the pipe line presents no particular difficulty, because the excavation once correctly started is not likely to deviate to any injurious extent. It is much more difficult, as it is more important, to keep the grade. This is best effected, practically, I think, in the following manner : Let the ordinary marked grade stakes be set for the excavation. Then when the proper depth has been reached, or nearly so, let grade plugs be driven in the bottom of the trench, every 50 feet or oftener, with their heads exactly to grade. . A line can then be stretched from one to the other, and the blocks laid to it. It is better to bed the blocks a trifle low, say a quarter to half an inch, particularly with heavy pipe and hard bottom, and then raise the pipe to grade by driving in the wedges. It is not necessary to set the pipes with rod and level. If the grade plugs have been driven as suggested, a competent foreman will adopt any one of many ways for sighting in the pipe to the proper level. With soft ground and heavy pipes, longitudinal stringers are advantageously employed under the blocks, the spaces between them being well packed with broken stone or other ballast.

When pipe have been laid and calked, it is advantageous to cover them as soon as possible by backfilling the trench, to prevent the joints from drawing in consequence of expansion and contraction due to exposure to the changes of temperature of the air.

In backfilling the trench after the pipe is laid, be very careful that the earth is well tamped in under the pipe, so that it may have a solid bearing throughout its entire length. The earth put in next to the pipe should be clean and free from stones. Be particularly careful that no large stone gets under the pipe, as in case of a sudden jar, such as would be produced by a casual "water hammer," it might punch a piece out of the pipe, or at least crack it.

In leading and calking joints, the specifications generally call for a certain depth of lead. The specifications of the city of New York require 4 ins. of lead for 48-in. pipe. It is a great advantage to have a deep joint, although the necessity is sometimes denied upon the ground that in calking it is impossible to "upset" the lead to a depth of more than perhaps half or $\frac{3}{4}$ of an inch, and that therefore the additional lead is of no advantage. This is, however, a mistake, for an abundant depth must be allowed for cutting off the protruding lead in recalking a joint which has drawn. The lead which is drawn out when a line of pipe contracts is not forced back again when the pipe expands; on the contrary it remains out, and, if the pipe contracts again, a fresh amount will probably be again drawn out, all of which must be cut off when the joint is recalked.

In calking joints, yarn or gasket is first driven in until the proper depth is secured, and lead is then poured in, either by the use of the clay dam or "snake," or by the use of a metallic "clip." The latter is very useful in laying large pipe, as it enable the whole joint to be run at one pouring. A refinement in running joints is to employ a lead gasket, for example a piece of lead pipe, hammered flat, and circled around the pipe. It is driven in and calked, and the remainder of the joint is then run and calked in the usual manner. The advantages of this system are that there is no perishable material used, and that the joint is calked on both faces, which is very favorable to making it tight. It is sometimes difficult to get the trench and pipe sufficiently dry to admit of pouring a molten joint. In such cases it is necessary to put the lead in cold, perhaps in the form of a ring of flattened lead pipe, as above, and calk as put in. This makes a very satisfactory joint, only it is slower and more expensive.

Mr. Billings, in his excellent treatise, "Some Details of Water-Works Construction," gives Mr. Dexter Brackett's formula for the average weight of lead in a joint about $2\frac{1}{4}$ in. deep, as follows:

$$l = 2d,$$

in which l = pounds of lead per joint, and d = inside diameter of pipe in inches. As the pipes are usually 12 ft. in length, the weight of lead per running foot equals one-sixth of diameter of pipe in inches. If a 4-in. joint is used, we would have

$$l = 3.2d,$$

and the weight per running foot under above assumption of length of pipe would be $\dfrac{d}{3.75}$, or, in round numbers, $\dfrac{d}{4}$.

APPENDIX.

As it is convenient in making estimates to have the correct weight of cast iron pipes of different diameters in handy form, I subjoin a table of weights of pipes made by the Warren Foundry, of Phillipsburg, N. J.:

TABLE SHOWING THICKNESS OF METAL AND WEIGHT PER LENGTH FOR DIFFERENT SIZES OF PIPE UNDER VARIOUS HEADS OF WATER.

Inside diam., in inches.	50 ft. head.		100 ft. head.		150 ft. head.		200 ft. head.		250 ft. head.		300 ft. head.	
	Thickness of metal.	Weight per length.	Thickness of metal.	Weight per length.	Thickness of metal.	Weight per length.	Thickness of metal.	Weight per length.	Thickness of metal.	Weight per length.	Thickness of metal.	Weight per length.
2	.294	63	.312	67½	.330	72	.348	76½	.306	81	.384	86
3	.314	114	.353	149	.362	153	.371	157	.380	161	.390	166
4	.361	197	.373	204	.385	211	.397	218	.409	226	.421	235
5	.378	254	.393	265	.408	275	.423	286	.438	298	.453	309
6	.393	315	.411	330	.429	345	.447	361	.465	377	.483	393
8	.422	445	.450	475	.474	502	.498	529	.522	557	.546	584
10	.459	600	.489	641	.519	682	.549	723	.579	766	.609	808
12	.491	768	.527	826	.563	885	.599	944	.635	1,004	.671	1,064
14	.524	952	.566	1,031	.608	1,111	.650	1,191	.692	1,272	.734	1,352
16	.560	1,215	.604	1,253	.652	1,360	.700	1,463	.748	1,568	.796	1,673
18	.589	1,370	.643	1,500	.697	1,630	.751	1,761	.805	1,894	.859	2,026
20	.622	1,603	.682	1,763	.742	1,924	.802	2,086	.862	2,248	.922	2,412
24	.687	2,120	.759	2,349	.831	2,580	.903	2,811	.975	3,045	1.047	3,279
30	.785	3,020	.875	3,376	.965	3,735	1.055	4,095	1.115	4,458	1.235	4,822
36	.882	4,070	.990	4,581	1.098	5,096	1.206	5,613	1.314	6,133	1.422	6,656
42	.980	5,265	1.106	5,958	1.232	6,657	1.358	7,360	1.484	8,070	1.610	8,804
48	1.078	6,616	1.222	7,521	1.366	8,431	1.510	9,340	1.654	10,269	1.798	11,195

All pipe cast vertically in dry sand, in lengths of 12 ft., except the 2-in., which are cast 9 ft. long.

The general formula for weight of cylindrical cast iron pipe of given thickness is:

$$W = 0.82 (D + T) T \times L, \qquad (1)$$

in which

W = weight in pounds.
D = inside diameter in inches.
T = thickness of metal in inches.
L = length in inches.

PRACTICAL HYDRAULIC FORMULÆ.   69

A convenient approximate formula for the weight per foot of cylindrical part of a cast iron pipe is :

$$W = 10 (D + T) T. \qquad (2)$$

In calculating the weight of cast iron pipe, they are always considered as being cylindrical, and eight inches of length are added as the equivalent of the hub or bell for all diameters. Thus a pipe measuring 12 feet over all, including the hub, would be considered, as regards weight, as a plain cylindrical pipe 12 feet 8 inches long. For instance, to calculate the weight of a 42-inch pipe, 12 feet over all and 0.980 inch thick, by the above approximate formula (2) we have:

$$W = 10 \times 12.67 \times 42.98 \times 0.98 = 5336.6 \text{ lbs}.$$

This is about 1⅜% in excess of weight given in Warren Foundry table.

If we wish to find the thickness, the length, diameter and weight being given, we have from (1):

$$T = \sqrt{\frac{W}{0.82L} + \frac{D^2}{4}} - \frac{D}{2} \qquad (3)$$

To find the proper thickness in inches corresponding to the head in feet, H, of pressure for a given diameter in inches, D, we have :

$$T = 0.00006 \, H D + 0.0133 \, D + 0.296.$$

The hydraulic engineer will find it both interesting and useful to work out concise formulæ covering frequently recurring cases, and enter them in his notebook for future reference. These should be carefully checked by testing them numerically, so that they can be used with confidence when wanted. For instance, it is a common practice to allow 100 United States gallons per capita to be consumed in 10 hours, in calculating the proper supply for a town. To find the diameter of a pipe to convey this amount we may use the approximate formula ;

$$D = \sqrt[5]{\frac{(0.0004 \times N)^2}{H}}.$$

in which

D = diameter in feet.
H = fall per 1,000.
N = number of population to be supplied.

Also, in estimating theoretical horse power necessary to raise a given volume of water to a certain height and in a certain time, we have

$$\text{HP.} = \frac{Q \times H}{8.82}$$

and

$$HP. = \frac{Q' \times H}{5.7},$$

in which

Q = cubic feet per second.
Q' = millions of U. S. gallons per 24 hours.
H = lift in feet.

In calculating the power necessary to pump water into a reservoir situated at certain horizontal and vertical distances from the pumps, and connected with them by a force main of given diameter, we must imagine the water to be raised to a certain elevation above the reservoir such that the difference of level between this elevation and that of the reservoir shall be sufficient to convey the required amount of water to the reservoir against the friction of the force main. Thus, suppose we wished to deliver 1 cubic foot per second through a 12-inch force main to a reservoir 100 feet above the pumps, and distant 10,000 feet from them. By our approximate formula (13) we know that this requires a fall of one foot in a thousand, so, as the distance is 10,000 feet, we need 10 feet to overcome the friction, and our pumps must therefore be able to raise the given volume a total height of 110 feet. This calls for 12.47 theoretical horsepower, by the formula just given.

SECOND PART.

NOTES ON WATER SUPPLY ENGINEERING.

QUALITY OF WATER.

The first question in regard to a water supply is, evidently, *quality*. The solution of this question, either generally or in any particular case, is by no means a simple nor any easy one, particularly since it is now fully recognized that chemical analysis, to which we naturally recur in such cases, is far from being a final criterion, but is only one element in a group of data from which we infer the probable quality of a given source of supply.

As animal refuse and the waste products of human industry are the principal sources of menace to a water supply, we commonly look for a high degree of purity in water drawn from thinly populated districts devoid of manufactories. Such districts, however, are not often to be met with in the vicinity of large towns, and, even when they are, we must expect a gradual encroachment upon them, from the natural growth of the neighborhood.

One excellent criterion of quality is the general health of the communities using the water. Also, the order of fishes which find their habitat in the streams of a given district. Little danger, for instance, would be apprehended from the use of a good trout stream.

Although all the fresh water used upon the earth reaches it

in the form of rain which is first drawn up by evaporation, chiefly from the sea, there are several different forms under which it presents itself for our use. The first broad classification of these forms would be that which divides the supply into *surface water* and *ground water*. By surface water is understood the water of lakes, ponds, rivers and streams, all water which in fact is collected directly from the surface of the earth; and by ground water, that which is derived from wells and filtering galleries, and from springs when taken at or near their source.

Each of these classes admits of much subdivision, but the differences will be principally those of degree, and not of kind.

For instance, we have the smaller streams, such as creeks, brooks, etc., and also the larger ones, rising to the dignity of rivers. While these certainly do present slightly different characters, still their main difference is one of size. Again, though the waters of lakes and ponds differ somewhat from each other, and from those of streams and rivers, still they are only the collected products of these latter, which they consequently greatly resemble.

Ground water, proceeding directly from the earth, offers more distinctive characteristics, shared, generally, by all its subclasses.

As regards the relative salubrity of water drawn from minor streams and that from large rivers, it would seem that they stood nearly upon a par, their principal difference, as already mentioned, being that of size. At first sight it would appear that the smaller streams, situated near the headwaters of the larger ones, or rivers, would possess a higher degree of purity from the fact of the water being collected from a comparatively wilder and more thinly populated district. This is not, however, of necessity, the case, because the river, although passing through a denser population, affords by its greater volume a greater degree of dilution. The true criterion in respect to this view of the question would seem to be the density of population per square mile of drainage era, together with the proximity of its center of gravity to the stream itself. The farther the bulk of the population is from the stream or river, the greater the chances of purification by natural filtration. Large rivers are apt to have towns directly on their banks, which drain all their sewage immediately into the stream.

Large rivers, being made up principally of the smaller streams, would appear to form the general average of all their feeders. It must be borne in mind, however, that besides the yield of the smaller streams the river is partly fed by direct surface wash from its immediate banks, thus imparting to it a somewhat modified character, distinct from that of the smaller brooks emptying into it, and causing the water of the large river to be, as a general rule, softer and warmer than that of its small feeders. Generally speaking, however, in the large river you get *all* the water and *all* the impurities, thus making, as already stated, a pretty fair general average of the whole, while of the smaller feeders some will have a greater and some a less degree of concentration of impurities than the average.

There is another point worthy of note as regards the relative quality of the water taken near the mouth of a great river, or from the smaller streams near its source. All impurities entering such streams or rivers have a greater chance of being exterminated by oxidation, by the lower forms of organic life and by fishes, the longer they remain exposed to these agencies. Hence near the mouth of long rivers we have a right to assume that many of the impurities which entered near the headwaters have been destroyed during their long passage toward the mouth. On the other hand, this prolonged sojourn has increased the probability of development of disease germs which have escaped actual destruction. The question then comes up: Had we better take our impurities and disease germs fresh or stale? And the answer would probably be: Fresh, if we must take them at all, and cannot trust to time for their destruction.

There is another point of difference between the two classes of streams, which, although possessing an engineering rather than a sanitary character, it may not be amiss to refer to here. In the case of the small streams a greater necessity will generally exist for storage, in order to secure a uniform supply, while gravity can be more often counted upon as a motive power than in the case of the large river, where storage is seldom needed, and where pumping is almost invariably necessary; a notable exception being that of Washington, D. C., where the water of the Potomac flows into the city by gravity.

Ground water may be drawn from shallow or deep seated wells—the latter often improperly called *artesian*—from galleries, or directly from springs at the point where they burst forth from the earth.

Shallow wells are supplied by the rain which falls and soaks into the ground in their immediate vicinity. In seasons of much rain the level of saturation is comparatively near the surface; in seasons of drought the level descends as the water gradually drains off to the nearest valley.

While an isolated shallow well may afford water of excellent quality and considerable relative coolness, such wells situated in towns and villages, or even when located near the cesspools of solitary dwellings, constitute what upon the whole must be considered the most objectionable supply in common use. Their hardness and saltness when compared with neighboring springs are a good indication of their relative contamination by human refuse. These qualities are observed to increase with time, and the growth of the village—a striking corroboration of what has been advanced above.

Deep seated wells, and springs, may be fed by rain falling on far distant points. The water from these wells is apt to be impregnated with earthy salts, and therefore to be hard, frequently to the extent of unfitness for domestic uses. The temperature is apt to be higher than that of shallow wells.

The supply from deep wells is more abundant and steady than from shallow ones, the volume of supply being more dependent upon depth than diameter ; indeed increased diameter only affords greater storage in any given case.

Ground water is frequently obtained from drains or filtering galleries, or lines of pipe with open joints, chiefly located near and parallel to and lower than rivers, which galleries intercept the water flowing through the ground toward the river, and which probably are also fed, to some extent, by the water of the river itself, leaching back to the drain, gallery or pipe line.

Water in considerable quantities is sometimes collected from springs, and conveyed away immediately as it bubbles up from the ground. The new water supply of Havana, Cuba, is a notable instance of this, where some four hundred springs, furnishing over five millions of cubic feet per 24 hours, have been collected about ten miles from the city, and the water conveyed in an aqueduct to a distributing reservoir, whence it is delivered to the city in cast-iron pipes. Such a supply seems likely to be the purest that can be obtained. Nevertheless, spring water is apt to be somewhat hard from the amount of earthy salts frequently held in solution.

It will be seen from the above that the question of the relative purity of different classes of water is a very complicated and uncertain one, not admitting of a general solution, but involving the consideration of a great number of special cases. Ordinarily the choice in any given instance is very limited, most towns having but few sources of supply to select from. The choice is ordinarily further controlled and limited by questions of quantity and cost, so that it seems hardly worth while to consider the subject under its general aspect at all, but simply to make a special study of each special case.

Quantity of Water.

Next in importance to *quality* comes the question of *quantity*. It will be observed that the growing tendency is to increase the amount allotted per capita per diem. It is found to be necessary to make abundant provision for the future growth of the town to be supplied, to anticipate an increasing individual use of water, and also to provide for the yearly increase of leakage, consequent upon the gradual deterioration of the work and of the house plumbing. This latter is a fruitful though often overlooked cause of a diminishing supply.

In general it may be said that a hundred gallons per twenty-four hours per capita, to be consumed in ten hours, with a liberal allowance for future growth of population, is a safe but not extravagant estimate. We frequently hear of a town finding that its water supply has become inadequate; we never hear of one suffering from too great a one. The control and diminution of waste are now occupying a great deal of attention, particularly in England, where the density of the population renders strict economy necessary. Frequently, no doubt, the best and cheapest way of increasing a deficient water supply would be to reduce waste, by the use of meters and other means for securing the co-operation of consumers.

From the purely engineering point of view the principal interest involved in the hourly supply is its connection with the size of pipes required for its delivery. A hundred gallons per head per twenty-four hours if delivered in ten hours, is at the rate of ten gallons per head per hour, or about 0.00037 cubic foot per second.

It has just been stated that quantity is secondary to quality, but in studying a water supply project the first step is to decide upon the quantity necessary or desirable to obtain. This fact being settled, the question will naturally follow, How can we ascertain what the yield of a given stream will be ? One way is by gaging, and this should be always done, choosing both the driest and wettest seasons for the purpose. But it is evident that all this takes time, and even a year's continuous gaging would not be considered as conclusive in any case where the demand nearly approaches the probable supply, because we must calculate on an occasional year or two of very exceptional drought. Another way is to make a survey of the area which drains into the stream under study, above the point at which it is proposed to take the water. This area, combined with the rainfall, known or assumed, and a general knowledge of the character of the watershed, furnish reliable data for calculating the approximate yield of the stream. Here again, however, we are confronted with the necessity of consuming much time, for although the survey can be rapidly made, the records of rainfall require at least as much time as does gaging. Fortunately in many cases we can make pretty close estimates of the amount of water probably derivable from a given area, by using data already collected for neighboring districts, and at any rate we can always make reasonable assumptions when once we know the number of square miles of territory which drain into our stream—the liability of such assumptions to be correct increasing with the area, for a large area is less subject to special variations from local causes than a small one.

The average yearly rainfall in the Croton basin, which furnishes by far the larger part of the water supply of the city of New York, is about 46 inches. Long experience shows that in this basin each square mile of watershed, or drainage area may be safely counted upon, one season with another, to furnish one million of U. S. gallons per twenty-four hours, or 365 millions per year. On the other hand, a precipitation of 46 inches gives very nearly 800 millions per square mile per year. Hence, in the Croton basin about 46% of the total rainfall is found to be available for water supply.

It must be borne in mind that the above yield represents the *yearly average*, which may easily vary forty times either way for any given shorter period. This fact establishes two important points in regard to water supply. First, the necessity of adequate storage reservoirs to convert this yearly average into a daily one; and secondly, the necessity, in the interest of safety, to give these reservoirs ample overflows, or spillways, in order to provide free escape to the surplus water which may flow into them in immense volumes during freshets.

These considerations bring us naturally to the question of storage, a most important and by no means simple one. The amount of storage necessary to insure a regular daily supply varies of course with the extent of the watershed in proportion to the demand. The larger the area, the smaller may be the storage. In some exceptional cases the supply may be so great that its absolute minimum yield is greater than the maximum demand, and in such cases no storage is necessary. On the other hand cases so unfavorable may possibly occur when the total yearly average is needed, and this leads to a maximum storage capacity.

Let us consider such a case, and suppose a community which requires a supply of 10 million gallons a day from a drainage area of 10 square miles, and follow the course of events through an entire year. The year will be divided into three periods: The period of average flow, the period of drought, and the period of over-supply.

The period of average flow will be that in which the daily yield of the stream is exactly equal to the daily draft—in our supposititious case 10,000,000 gallons. The period of drought will be that in which the yield is less than the above, and the period of over-supply, that in which it exceeds it. These last two periods will, of course, vary in intensity, through indefinite gradations.

In order that the storage capacity should be ideally perfect, it would be necessary to so proportion it that at the commencement of the period of drought the reservoir should be exactly full and of capacity sufficient to bridge over the interval between the drought and the commencement of the period of average yield. At the commencement of the over-supply or freshet period, the reservoir should be completely empty, and of capacity sufficient to receive and retain all the surplus water until the period of average flow was again reached. Not a drop should ordinarily escape except through the supply pipes, and an overflow or spillway should be unnecessary except to provide for extraordinary contingencies, such as cloudbursts, etc.

It is clear that such an ideal state of things is impossible of realization. It would be based upon a regularity of regimen that could never occur except, perhaps, by chance, during a single year. Even in periods of extreme drought (and this extreme is a variable) there would be some water flowing in the stream, and the storage reserve would need to be drawn upon only for the difference between this amount and the daily supply; while during freshets the amount to be stored would be reduced by the daily supply being drawn off. Moreover, besides the average intensity of droughts and freshets, there come cycles of still greater intensity, all of which circumstances are controlling factors in the problem.

In the case assumed the only way to secure the total flow, so that none shall pass to waste over the spillway except in the case of a cloudburst or of some other phenomenon, and at the same time to provide for droughts of maximum intensity, is to construct a reservoir or reservoirs of capacity to contain the total yearly yield of the stream, and to commence the use of the supply with a full reservoir, so that there shall always be a year's supply ahead.

This treatment of the problem is certainly a heroic one, and has probably never been fully carried out in practice, although the city of New York is reaching well on toward it, in the vast storage works executed and contemplated in the Croton basin. Fortunately so unfavorable a case as the one assumed, when the total yield of the stream is needed, rarely presents itself, and in the majority of cases there is an excess, more or less considerable, of the supply over the demand.

The solution of the problem of storage capacity lies between the two extremes above instanced, in one of which no storage is necessary, and, in the other, when it is necessary to have capacity for the yield of the whole year.

It is evident that we cannot say, *à priori*, of any proposed water supply, that storage for so many days will be necessary, or sufficient, without knowing at least approximately the total yield as well as the desired consumption. In cases where close calculation is needed, as when it appears necessary to utilize the greater part of the supposed supply, the proper course to pursue is to ascertain, by actual survey, the drainage area; to ascertain, by rational assumption when direct observations are lacking, the average precipitation, and then allow from one-quarter to one-third of the same as available, backing these data by gagings, as complete as may be possible, of the stream, and calculate the storage capacity accordingly.

I have confined myself in the above to a general view of the principles involved in planning storage reservoirs, nor do I think it wise to enter into more elaborate calculations, as they might lead the inexperienced to suppose that the problem really admitted of a general mathematical solution. Such is not the case, and unless the known factors point clearly to self-evident assumptions, great caution and much study should be bestowed upon the fixing of the data on which the design of an economical and satisfactory water supply is to be based. One thing is certain : Except for economical reasons there is no danger of having too great storage capacity. I do not happen to recall an instance of a community suffering from the possession of too much stored water, while the want of enough of it is proving a serious trouble to cities and towns all over the country.

I have already adverted to the matter of adequate spillways for discharging the floodwaters of freshets. There is no uniformity of practice for the dimensions of these all-important adjuncts, and it is probable that the great majority of those now in existence have been proportioned by guesswork, or, "upon general principles." This is all wrong; and in this, as in all other questions of design, we should first ascertain what conditions our structure will have to fulfill, and then dimension it accordingly.

The capacity or open area of a spillway, is made up of its length and height of notch. It must be large enough to pass all the water of extreme floods without danger of over-topping the dam. Forty times the average flow, or 40 million of gallons per square mile and per 24 hours—or 64 cubic feet per square mile per second—is none too liberal an allowance, particularly for earthen dams. For this amount of water we have the two simple approximate formulæ to determine the length and depth of notch of a spillway, the depth being counted from the level of the lip of the dam to the surface of still water in the reservoir:

$$L = 20 \sqrt{A}, \quad (1)$$
$$D = {}^3\!\sqrt{A} + C \quad (2)$$

in which $L =$ length in feet, $D =$ depth in feet, $A =$ area of watershed in square miles, and $C =$ a certain additional height above the water in the reservoir, depending upon the character and construction of the dam. If we should wish to provide for a different amount of water, we must generalize formula (2), writing:

$$D = \frac{{}^3\!\sqrt{Q^2}}{16} \times {}^3\!\sqrt{A} + C. \quad (3)$$

in which $Q =$ cubic feet per second per square mile.

DAMS.

Large reservoirs are generally formed by building a dam across the valley of the stream furnishing the supply. Naturally, the narrowest point is chosen, but further investigation may prove such point to be not the most favorable one. A solid foundation is the first requisite, and sometimes firm rock is found so much nearer the surface at a point where the valley is wider that a dam built there would be actually shorter than at the narrower point, besides saving the extra excavation to get down to solid bottom. It is abundantly worth while to devote considerable time in exploring and surveying, before fixing definitely upon the location of the proposed dam.

In examining the character of the foundations, I think that test pits furnish the only trustworthy information. At great depths, these would be very expensive, and recourse is generally had to drilling. This furnishes good indications when properly interpreted, but also has occasioned many expensive misconceptions of the ground. The test pit remains the only sure means of ascertaining what is below the surface.

The character of the ground will determine the class of dam which should be built. If good rock bottom is to be found, a masonry dam will be the best, and perhaps not much more expensive than a properly constructed earthen one. All the elements of a masonry dam are more fixed and precise than those of an earthen dam can be, so there is less necessity for piling up what may in reality be redundant work, to provide for contingencies which we cannot exactly determine quantitively.

Masonry dams may be divided into three classes; low, medium and high. Although the lines of demarcation are somewhat vague, we may class all dams less than thirty feet high as low, those between thirty and sixty as medium, and all those above sixty as high. Before commencing our investigations it will be extremely useful to establish certain data.

Calculation shows that the equation of equilibrium of a dam with vertical faces is:

$$Wx^2 = 20.83 H^3, \qquad (1)$$

in which $W=$ weight in pounds of a cubic foot of the masonry, $H=$ the height of wall, and $x=$ its thickness, both in feet. The weight of a cubic foot of water is assumed at 62.50 pounds. From this equation we derive :

$$x = \frac{4.565\,H}{\sqrt{W}}. \qquad (5)$$

These equations show that the overturning moment varies as the square of the height, and the resisting moment as the square of the thickness, and the square root of the density of the masonry; while the value of x, the thickness, varies as the height, and inversely as the square root of the density, of the wall ; that is to say, from (5) we deduce :

$$\frac{x\sqrt{W}}{H} = 4.565,$$

a constant.

If we assign 125 pounds as the unit weight, or weight per cubic foot, of the masonry, we find :

$$x = 0.41\,H. \qquad (6)$$

This is the value of x for exact static equilibrium. We may obtain whatever factor of safety we wish by simply multiplying the square of 0.41 by such factor and extracting the square root of the product. Thus, suppose we wish a factor of 2.5. Operating as above, we find :

$$x = 0.648\,H. \qquad (7)$$

As the assumption of weight is somewhat arbitrary, we may for simplicity write (7) thus :

$$x = \frac{2H}{3}; \qquad (8)$$

that is to say, a "plumb" wall to resist water pressure should be twice as thick as one to resist average earth pressure.

However, dams are not built plumb. They generally have vertical backs, toward the water, and battering faces. The readiest way to transform a vertical or plumb wall into a trapezoidal one of equivalent resisting moment is to follow Vauban's principle, that all equivalent walls with vertical backs have the same thickness at one-ninth of their height from the bottom. This rule holds very closely good within wide limits.

As an application, let us take the case of a wall to sustain water, 27 feet high. If vertical, its thickness should be 18 feet for a factor of safety of 2.50, and we would have the rectangle $ABCD$, shown in Fig. 16. Transforming, according to Vau-

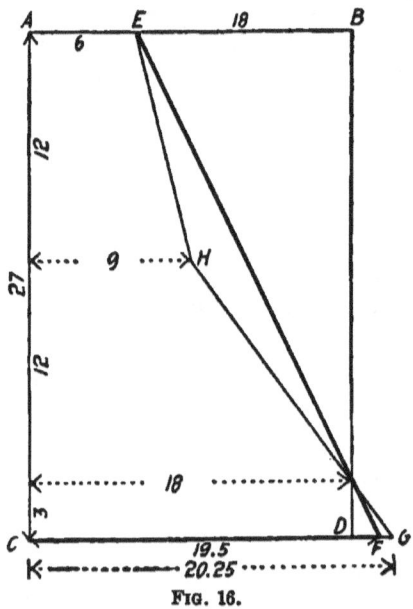

Fig. 16.

ban's rule, to a trapezoidal section, with top width of 6 feet, we get the figure $AEFC$, of which the bottom width CF is 19.50 feet. Verifying the comparative stability of the two sections, we find that of the trapezoidal one to be within about 1½% of the rectangle, while its area is nearly 30% smaller.

Such a section as $AEFC$ is obviously awkward, presenting a top-heavy appearance, from the redundant thickness of its upper half. Some such section as $AEHGC$ would be preferable; it has still less area and not much less stability than the trapezoid $A\ E\ F\ C$.

Passing to dams of medium height, let us take for an example one 54 feet high (Fig. 17). The proper thickness, if vertical, per

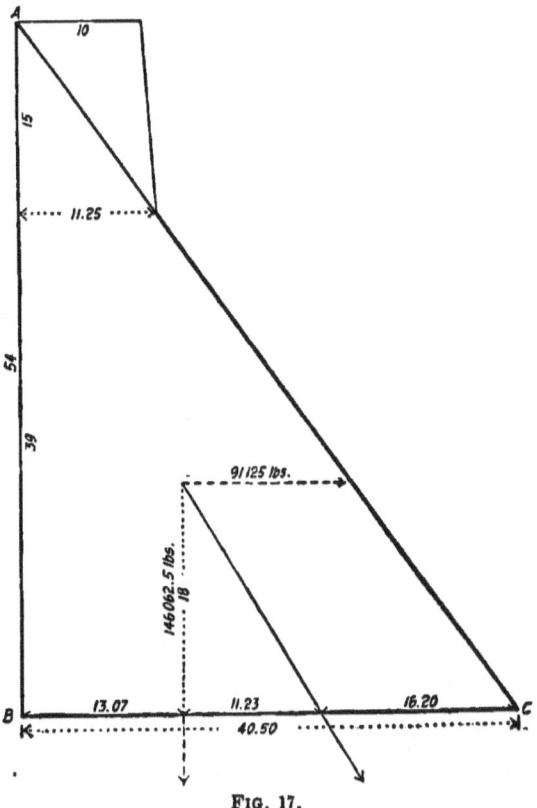

FIG. 17.

formula (8) is 36 feet. Transforming the rectangular section into a *triangular* one, by Vauban's rule, we obtain the right-angled triangle $A\ B\ C$ which possesses certain interesting properties. In the first place, the triangular section has only about 85% of the stability of the rectangle. Secondly, its base will always be three-quarters of its height. Thirdly, the resultant of its weight combined with the thrust of the water (still assuming the specific gravity of the masonry to be 2) will always cut the base about 11% within its "middle third"; while, of course, the action of the weight alone will cut the base exactly at the inner extremity of the middle third.

Evidently such a section is impossible in practice, because it involves a top width of zero. Let us give it a top width of 10 feet, with a face batter of an inch to the foot to the upper part. This batter will always intersect the hypotenuse of the triangle at a distance from the top equal to one and a half times the top width, whatever that may be.

This composite structure has a resisting moment about 8% less than that of the rectangular one of the same height, and base of 36 feet. It still has a factor of safety of 2.42, with the above relative densities of masonry and water, while the section is about 40% less than the rectangular one. Fig. 17 shows all dimensions, and the triangle of forces. It will be noted how the addition of the upper trapezoid modifies the points of application of the pressures.

In regard to all dams, high or low, we may lay down the leading principle that the line of pressure should always pass within the middle third of the base, especially the line which corresponds to a full reservoir; that is, the line which is the resultant between the weight of the dam itself and the thrust of the water. In very high dams it is not sufficient that this condition be fulfilled for the base only: it must hold good also for any horizontal bed between the base and the top, because in such dams, in order to economize material, the face is given the form of a convex curve, and if this convexity be too great it will occur that, while the base may have a satisfactory width, some of the upper beds parallel to it will not. The object in designing a high dam is to give the section such a form that it shall be a "section of equal resistance," because this is always the section of greatest economy.

The problem is further complicated in the case of very high dams by the fact that the resistance to overturning is not the only thing to be considered. We must also determine whether the area of the lower beds is sufficient to resist the crushing strain brought upon them by the weight of the superincumbent mass. In making this investigation it is obvious that the first step will be to fix a proper limiting unit strain, or admissible pressure, per square foot upon the masonry. This limit depends upon the nature of the material used and also upon the views of the designer. For ordinarily good masonry, 15,000 pounds per square foot would be considered a conservative limit, being a trifle over 100 pounds per square inch. If the resultant of the pressures cut any bed exactly in the middle, we could ascertain the pressure per square foot upon such bed by simply dividing the whole weight of the mass resting upon it by its length. But when the resultant moves from the center, the strain is no longer evenly distributed over the entire bed, but is intensified upon that portion comprised between the point where the resultant cuts the bed and the nearer extremity of the same, reaching its maximum intensity at the extremity itself, or as we should say, at the *nearer toe*. The investigation of this varying strain, which increases in proportion as the resultant approaches the nearer toe, is somewhat obscure, and rests upon assumptions of somewhat unsatisfactory demonstration. The following two formulas may, however, be accepted as reliable approximations to the truth, within the limits occurring in ordinary practice:

$$P = \frac{2W}{3D}, \qquad (9)$$

$$P = \frac{4W}{L^2}(L - 1.5D), \qquad (10)$$

in which:

$P =$ pressure, in pounds, per square foot.

L = length of given bed, in feet.
W = weight, in pounds, of mass above given bed.
D = distance, in feet, from point of intersection of resultant with given bed, to nearer extremity of same.

Formula (9) is used when D is equal to or less than $\dfrac{L}{3}$. Formula (10) is used when D is equal to or greater than $\dfrac{L}{3}$. When $D = \dfrac{L}{3}$, either formula gives unit strains equal to twice the total weight above bed, divided by its length.

We have then the three following conditions which the proper section of a high masonry dam should fulfill: *First*, the lines of pressure should lie within the middle third of all beds. *Secondly*, the maximum unit strains should not exceed a moderate fixed limit. *Thirdly*, the section should be one of equal or nearly equal resistance.

Now then, in the light of what has been already established, let us feel our way toward the proper design for a dam 160 feet high fulfilling the above three conditions. Let us assume a density of masonry double that of water, a limiting unit strain of 15,000 pounds, and, as is usual in such cases, let us consider a length of one foot of dam, so that the area of our section in square feet will represent an equal volume in cubic feet.

Knowing that one of the necessary conditions is that the resultants shall lie within the middle third of all the horizontal beds which we may suppose to divide the section, we feel sure that we cannot go far wrong in first laying down the right-angled triangle $A\,B\,C$ (Fig. 18), of base equal to three-quarters of the height, or 120 feet for the total height of 160 feet. Desiring a top width of say 20 feet, we lay off the same from A, giving the face of this portion of the section a batter of $\tfrac{1}{12}$. This batter we already know will strike the hypotenuse 30 feet vertical from the top. Now as we know that the effect of placing this top story upon our triangle will be to draw the line of vertical downward pressure, due to weight of masonry alone, away from the center of gravity of the triangle *A B C*, and therefore outside of the "middle thirds" on the water side; and, also, anticipating a

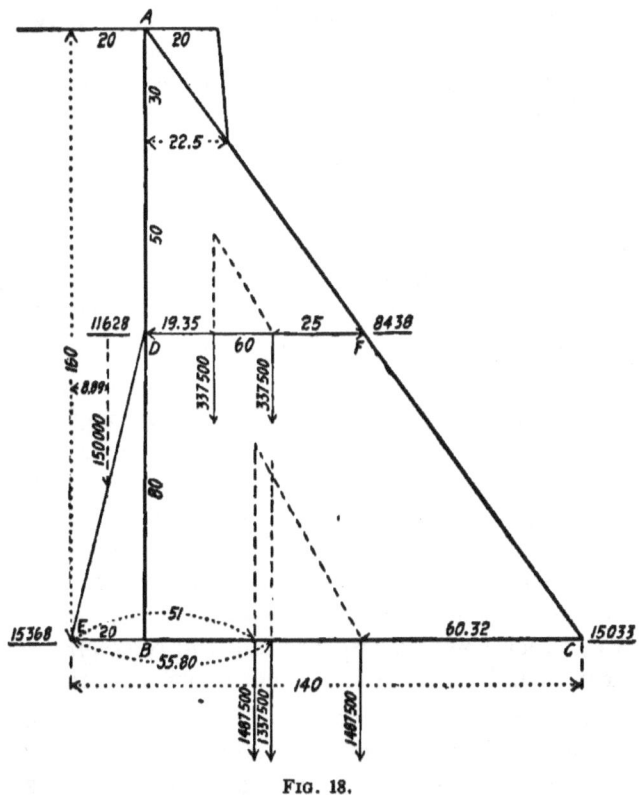

Fig. 18.

little the knowledge which we shall presently acquire, we give the back of the section, from a point 80 feet from the top, an outward flare of one to four, which is shown in the figure by the small triangle *D E B*, and which increases the total width of base to 140 feet. The section is thus divided into three trapezoids, respectively 30, 50 and 80 feet high, with corresponding widths of 20, 22.5, 60 and 140 feet.

Next we determine, either graphically or by simple calculation, based upon the properties of similar triangles, the points where the line passing through the center of gravity of the superincumbent mass cuts the imaginary bed $D\ F$, and also where it cuts the same when shoved forward by the thrust of the water, acting at right angles to it. These points are shown in the figure to be respectively at 19.35 ft. and 25 ft. from D and F. The figure shows also the intensity of the strain in pounds at these points, obtained by multiplying the total area above $D\ F$ by 125, the assumed weight in pounds per cubic foot of masonry, although the calculations were actually made in units of volume, for the sake of rapidity and ease, counting the weight of a cubic foot of masonry as 1, and that of water as $\frac{1}{2}$.

We next proceed in the same way in regard to $E\ C$. Here we have first the point where the line passing through the center of gravity of the entire section $A\ E\ C$ cuts the base $E\ C$, 55.80 ft. from E, and which corresponds to an empty reservoir; and secondly, the point, 60.32 ft. from C, where the resultant of the weight of the section $A\ E\ C$, plus the weight of water resting upon the inclined surface $D\ E$, combined with the forward thrust of the water acting under a head of 160 ft., cuts the same base $E\ C$, which point corresponds to a full reservoir. The intensities in pounds of all these strains are shown on the figure.

Our design now fulfills one of the imposed conditions: The lines of pressure lie well within the middle third, except at D, where the condition is not so binding. It remains to see how it complies with the second one. For this, we recur to our formulæ (9) and (10); and first to ascertain the unit strain at D. For this we employ (9), within which the case just falls. Substituting numerical values, we have :

$$P = \frac{2 \times 337500}{3 \times 19.35} = 11628 \text{ lbs. per sq. ft.}$$

For the strain at F we use (10):

$$P = \frac{4 \times 337500}{3600}(60 - 37.5) = 8438 \text{ lbs. per sq. ft.}$$

Passing to EC, we have for maximum strain at E, when reservoir is empty,

$$P = \frac{4 \times 1337500}{19600}(140 - 83.70) = 15368 \text{ lbs. per sq. ft.}$$

For maximum strain at C, when reservoir is full:

$$P = \frac{4 \times 1487500}{19600}(140 - 90.48) = 15033 \text{ lbs. per sq. ft.}$$

The maximum strains are given on the plan by the underlined figures at D, F, E and C.

Examining our design, we see that although it practically satisfies the first two conditions demanded, it is by no means a section of equal resistance, for the strains at D and F are far less than those at E and C. Evidently the upper bed DF is too wide.

As a further step in our tentative process, I will now offer, Fig. 19, a section suggested by Señor D. E. Boix, in his excellent treatise on "*La estabilidad de las construcciones de Mamposteria,*" as a general approximate type for high masonry dams. Beginning at the top, the skeleton of this design consists of a right-angled triangle ABC of base equal to two-thirds of its height, which height Señor Boix makes a constant of 24 meters, or say 80 ft. From B the back slopes to D with a batter of $\frac{1}{2}$, and from C to E with one of $\frac{3}{4}$. The skeleton of the design, therefore, in this particular instance of a total height of 160 ft., consists of a trapezoid $BCDE$ 80 ft. high, 140 ft. wide at bottom and 53.33 at top, surmounted by a right-angled triangle 80 ft. high. This upper triangular part is a constant, for all sections; the variation occurring in the lower trapezoidal part, according to the total height of dam. As a practical detail, the upper part is surmounted with another triangle, giving the section a proper top width. I have assumed a top width of 20 ft., with batter $\frac{1}{12}$, to correspond with previous example. These dimensions, with the triangles of forces and strains per square foot at the points B, C, D and E, are shown in the figure.

This may be considered an improvement upon the previous design. Its area is about 5% less and there is a much better distribution of strains. Its resisting moment is a little less, being as 2.87 to

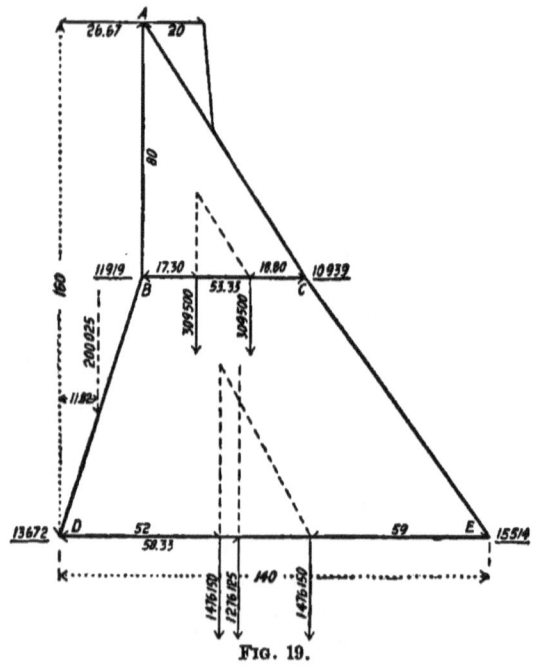

Fig. 19.

3.10, but the coefficient of stability is sufficient. Its practical superiority lies in the 5% of economy. In a careful final study of a high dam, it would be necessary to pass a greater number of horizontal beds through the section, and calculate the unit strains at each extremity of each. The result would probably lead to a more pronounced curve on the face, with a corresponding saving of material.*

---

* It is well, however, that the upper part of the dam should have a greater proportional strength than the base, because it is exposed to a greater degree of wave action.

An example of a well-proportioned dam, 80 feet high, is given, with all its dimensions and the triangles of forces, in Fig. 20. Calculations, similar to those used already, show maximum strains to be as follows: At E, empty reservoir, 8,677 pounds per square

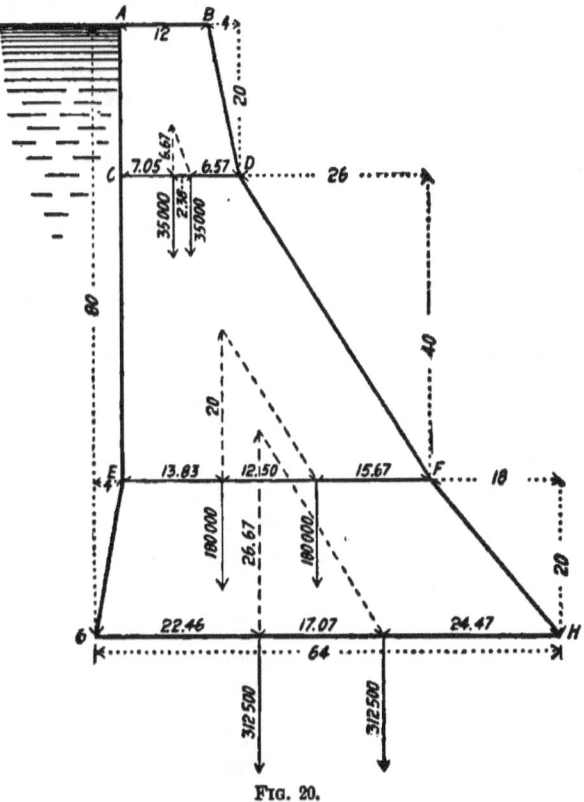

FIG. 20.

foot; at F, full reservoir, 7,607. At G, empty reservoir, 9,250 pounds per square foot; at H, full reservoir, 8,334. The water pressure on EG has not been counted.

In designing a high dam, it should be borne in mind that too much confidence is not to be placed upon the formulæ used in calculating the unequally distributed strains, and that the further their resultant moves from the center of the beds the less they are to be depended upon. As a further complication Señor Boix, in the work already mentioned, calls attention to the fact that, instead of considering only the vertical component of the resultant acting upon a horizontal bed, we should consider the resultant itself acting upon an imaginary bed inclined at right angles to it. He shows that this gives a more or less augmented unit strain. In an example which he gives of a dam about 100 feet high, the pressures thus calculated exceed those calculated upon the hypothesis of vertical action upon a horizontal bed by 10% to 17%. It would greatly, and I think unnecessarily, complicate the problem to treat it in this manner, for the weight of the superincumbent mass of masonry and the angle of the resultant are interdependent, and tedious processes by trial and error would be needed for each bed. The circumstance is merely referred to in order to show the necessity of keeping well within the margin of safety, for it must be borne in mind that a dam once built cannot readily be remodeled, and should stand intact and without material repairs as long as the town does which depends upon it for its water supply. The prototype of the modern high masonry dam is that across the valley of the Furens, near St. Etienne, in France. The perfect success of this dam is no proof that its section is suited for one of indefinite length, for it is itself very short, and wedged in between the rocky sides of the narrow valley which it spans, thereby receiving great additional strength from these lateral supports.

As regards the plan of a high masonry dam—that is, whether it should be straight or curved, with the convexity up stream—it cannot be said that a curved plan is necessary, nor, on the other hand, can it be denied that such plan is an element of strength, particularly if the dam be short. By adopting this form, in case of a slight movement occurring when the dam comes to its bearings under pressure, the character of the strain will be always compressive; while, if the axis be straight, any forward movement, however slight, will produce tensile strains, which are always to be avoided in unelastic materials like masonry.

Concerning the further details of the design of high masonry dams, they should stand upon a base or pedestal of which the top is level, more or less, with the surface of the ground, with ample offsets or projections beyond the toes of the dam proper, particularly on the downstream side. The excavation should go through the superincumbent earth, to and into the solid rock. The sides of the base should be vertical and, in the rock, built or packed close against the sides of the excavation. In earth, the space between the vertical sides of the base and those of the excavation should be filled up solid with closely compacted and puddled material, so as to leave no vacancies. The angle which the outer slope of the dam makes with the horizon should never be less than 45°, so as to avoid sharp and weak edges at the points of maximum strain. Whenever sufficient material can be obtained from the excavation or from borrow pits, a sloping bank of well-compacted earth should be placed against the back of the dam for at least a quarter or a third of its height from the ground, and protected by riprapping. This bank has a double object: it not only impedes leakage, but by establishing a permanent thrust, or at least support, against the back of the dam when the reservoir is empty, it diminishes the range of pressures existing between a full and an empty reservoir. All of these features are shown in Fig. 21.

Before leaving this question of design, there is a recommendation which I think it well to make, not only in regard to high masonry dams, but to all other engineering structures as well. It is this: After carefully proportioning the work according to approved methods of calculation, *take a good look at the drawing,* and see if it LOOKS right; if not, there is probably something wrong in the calculations, and they had better be gone over to see where the mistake is.

Where rock foundation is not obtainable, the best kind of dam will be an earthen one, with masonry core or center wall. No earthen dam can be considered safe that is not provided with such wall, carried down to a water-tight or comparatively water-tight stratum. The masonry center wall is the only sure and permanent means of cutting off percolations through the bank, and for this reason it should be carried well down, on the principle that the worse the character of the ground, the deeper should

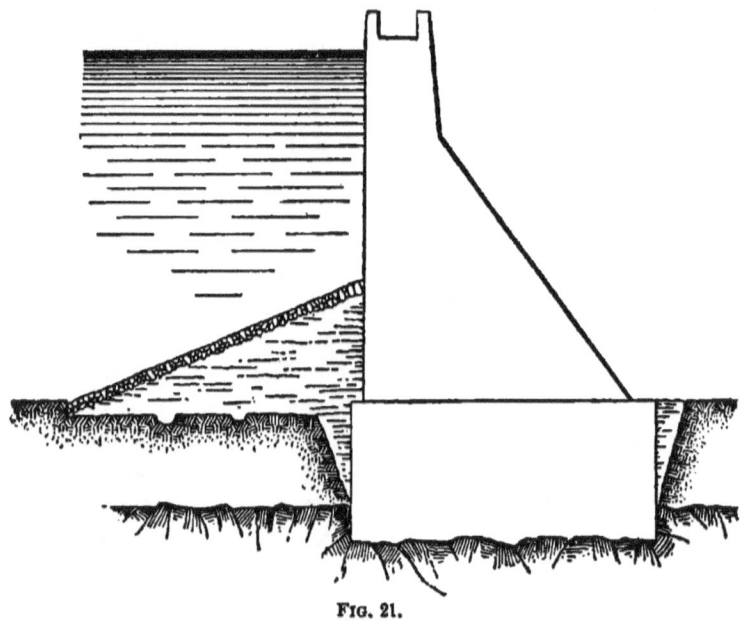

FIG. 21.

be the foundations. It should also be deeply imbedded in the sides of the valley. The worst kind of ground is a loose drift formation, containing many large cobblestones. In such material the wall must be sent " 'way down," till perhaps there is as much under as above ground, and the embankment, particularly on the water side, carried far back with a very easy slope, sometimes as much as five or six to one. Clay, sand, and fine, compact gravel are the best bottoms to build on; quicksand is also excellent, when lateral escape can be prevented, as it usually can be except when very near the surface.

Not only does the masonry center wall prevent percolation, but it also affords the means of making perfectly secure connections with all the accessories of the reservoir, such as culverts, piping, gate towers, etc. The failure of many earthen dams has been due to water following along outside the culverts or pipes by which water was drawn from the reservoir, a circumstance which cannot occur when these appliances are bonded in with a tight or even partially tight masonry wall, extending from flank to flank of the valley, and carried down to a good bottom.

Although this wall is supported on both sides by the embankment, it should be of sufficient thickness to withstand any unbalanced pressures to which it may be subjected. Where highest it may have a bottom thickness of about one-quarter its height, and be drawn in at the rate of about an inch to the foot, but preferably by offsets rather than batter. This would always give a top width equal to one-twelfth the height, which might sometimes lead to unsuitable dimensions, requiring modification. The top should be carried up at least as high as the level of highest water in the reservoir.

The spillway should be proportioned according to the principles already laid down, and if possible some natural depression should be found back of the dam, by which the overflow may be passed over to another valley, or into the same one, lower down. Failing this condition, a massive masonry spillway and apron must be built, generally in the axis of the stream, either with curved face, like the Croton or Bronx River dams, or stepped, like those of the Scranton Gas & Water Co., at Scranton, Pa. I am inclined to prefer this latter mode on its engineering merits, and apart from the fact of its greater cheapness. The section of the spillway will be proportioned according to the principles already laid down for high dams, but it should be even more massive, for it has to stand the impact of falling water. Its width at its foot should be equal to its height.

In regard to the embankment, it is ordinarily specified that this should be made of selected material. This is certainly recommendable, and the best material the site affords should be used in preference, but it is almost impossible to compel contractors to *sort* the material, and moreover really good stuff is not always within reach, in which case the volume put in the embankment must be increased, to compensate in a measure for its lack of quality. The surface of the bank, on the water side, should be well riprapped, and the lower slope sodded or seeded to grass. The bank should be kept wet by sprinkling while being put in, and should be brought up in horizontal lifts, and not made from a dump, like a railroad embankment. It should be compacted as it goes in ; ordinarily the travel of the carts, wagons and scrapers used will be sufficient for this purpose in connection with thorough sprinkling. The area covered by the embankment, particularly on the water side, should be carefully "floated off," grubbed and plowed, so that the material of the embankment shall come in contact with clean earth. It is also well to dig cross ditches, parallel to the center wall, to secure a still further bonding of the surface of the ground with the embankment. These ditches must be carefully filled and tamped with the material used in the embankment. An excellent form for the inside or water side of an embankment is a series of slopes and berms, forming one or more, terraces. Working from berm to berm gives a good opportunity to carry the work up level.

The best way to draw the water from the reservoir is by means of cast iron piping running through the center wall and terminating on the water side in a tower, built in with the center wall, and containing grooves for the reception of stop plank. The pipes should lead on the land side into an easily accessible gatehouse, also built in with the center wall, and each line should be provided with two good gates or valves, the inner one to be kept always open or partly open, and the outer one used for current operations. Then, should the latter get out of order, the inner gate can be closed, and the necessary examination and repairs effected. It is a good plan to build iron eye-beams into the walls of the gatehouse directly over the gates, so that in case of wishing to remove them the necessary overhead purchase can readily be applied. These ideas are not hypothetical, but have been satisfactorily carried out in the system of storage reservoirs built for the Scranton Gas & Water Company, already referred to.

It will now be well to make some remarks respecting the various classes of work embraced in the construction of dams and reservoirs. In such structures a good deal of concrete is generally employed, especially in foundations and subfoundations, for which purpose it is the best material that can be used. Great care, however, must be taken in its preparation and placing, for there is perhaps a greater difference between good and bad in concrete than in any other kind of masonry. Various proportions have been recommended, and we may lay down as a general principle, that the smaller the stones and the larger the percentage of cement used, the more water-tight will be the resulting concrete. I have obtained excellent results using fine ground Portland cement and thorough mixing, with one of cement, three of clean sand, and six of broken stone. In important work I should not care to use a less rich mixture than this, which, as I may add in passing, was that employed by the late General Gilmore in the rebuilding of Forts Sumter and Moultrie, in Charleston Harbor, with the exception of substituting Rosendale for Portland cement. A sufficient quantity of water should be added to bring the mass to a very decided degree of moisture, while not drowning it with an excess.

The sand and cement should be thoroughly mixed dry; then the stones, previously wetted, are added and the whole mass thoroughly mixed, water being added from time to time till it is brought to the proper consistency. Always keep the bed thin and avoid getting the material into heaps, which prevents proper mixing. If a mixing machine is used, the sand and cement should still be mixed dry by hand before putting into the machine. If the whole process is done by hand, there should be turned out about two cubic yards of concrete in the work per day of ten hours, per man all told, including all employed in mixing, placing and tamping.

This is about as much yardage as an ordinary gang of stone masons and helpers will do, laying up first class hydraulic rubble, and one of the principal reasons for concrete being cheaper than rubble is the fact of its being done by a cheaper class of labor. Indeed contractors generally bid entirely too low on hand made concrete. as compared with rubble, with the result that they endeavor afterward to do as little of it as possible, trying to get rubble substituted for it. Tamping should be carried on until the mass assumes a jelly like consistency, or at least till moisture is brought to the surface. This is a *sine qua non*, which inspectors must insist on. Prolonged tamping will frequently bring up the water from a piece of concrete which appeared to be hopelessly dry when put in. Bear in mind, that although too much water is bad, too little is very much worse. As a general rule the greater the percentage of broken stone, the more thorough must be the mixing and tamping. If the stone be hard and sharp the concrete will gain by giving it all that it will carry. Concrete may very easily be made too rich in mortar by carrying out the erroneous idea that the smaller the percentage of stone the better. After concrete has been placed, let it remain undisturbed and be kept moist by constant sprinkling for as long a time as possible.

The stone masonry used in the construction of hydraulic work is special in its character, and requires therefor special care in its execution. This fact should be fully explained to prospective bidders before they send in their proposals, or there will surely be remonstrances when the work begins. One principal source of contention is in regard to the size of the stones used. The work goes on so very much quicker by the use of the largest stones the derrick can lift, that both engineers and contractors are tempted to get in as many as possible. I think, however, that the fact remains the same that a more water-tight wall can be built with small stones. "Small stones" must be taken in a relative, and not an absolute, sense in this connection; near the foot of a thick wall larger ones can safely be used than in the thinner upper portions; and in hauling and distributing stone on the ground, contractors should be warned to keep the smaller ones handy for the top.

Great care must be taken in bedding and jointing the stones. They should be laid on the natural bed, the best and flattest bed down, and should be made to *swim* on their beds; *i. e.*, they should be susceptible of being swayed from side to side with a bar after they are laid and before being spalled up. No stone should rock when stepped on. When work first begins, stones, after being bedded and pronounced all right by the foreman, should frequently be raised again, so that he may see that, after all, there are many blank patches on the bed of the stone where it did not come in contact with the mortar. All stones, particularly small ones, should be hammered down on their beds. One advantage of using large stones is that they bed themselves to a great extent by their own weight. All the joints should be carefully made up, the invariable rule being, throughout the wall, that there shall be no vacant spaces, but that all that is not stone must be mortar. To this end, stone should not be laid too close together, but ample space afforded for getting the trowel all around them so as to cut the mortar well under and into the beds and joints. When the joints are narrow, they should be filled with mortar, and spalls driven in, as many as the joint will hold, without crowding, the only limit being that in no case shall there be stone to stone. Also be very careful that after a stone has been well bedded it shall not be lifted from its bed by wedging in too many big spalls under it. One of the tests of a good mason is the way he makes up his joints, fitting in all the spalls he can. Ordinarily, the mason finds it less trouble to fill up the joints with loosely thrown in mortar. This is against the interests of both contractor and company. Needless to say that the work must be well bonded, breaking beds as well as joints. Particular attention must be paid to the mortar, seeing that the sand and cement be thoroughly mixed dry, till it presents a uniform color, without streaks of sand and cement. Good average proportions are 1 cement and 2 sand for Rosendale, and 1 cement and 3 sand for Portland. The greater the proportion of sand the more thorough must be the mixing. Guard against having the mortar too wet, but let it always be worked up with the trowel to a soft pudding in the work. Masons frequently call out that the mortar is too dry, when what it wants is simply more tempering to bring out the moisture. The mason should be constantly tempering up his mortar, and when he has large quantities on hand he must call on his helper to do the same with his shovel. The more you work up and turn over fresh mortar—particularly when Portland is used—the better the results. As to quantity of wall laid up, per derrick, I think that each double drum steam derrick of proper sweep, well tended and constantly fed with materials, should be good for from 30 to 60 or even more cubic yards per day of ten hours. These limits are rather elastic, but perhaps it would not be fair to draw them closer, so much depending upon the size and character of the stones used. I have frequently timed a derrick, to see the number of trips made in a given time, but should be loth to fix a standard. If we allow five minutes to a trip, we should have one hundred and twenty per day. These trips convey not only the larger stones to be set by the derrick in the work, but also all the mortar and spalls needed for bedding and jointing. The governing factor, I think, will prove to be the size of the stones set by the derrick, for it will set a large one about as quick as a small one. One thing the contractor should closely watch in his own interest, and that is, that a derrick should never stand idle. If it is not constantly in motion he should at once find out what is the matter. Probably it is under-manned at one end or the other, and the gangs need to be increased. Strange to say, in regard to this question of rapidity, that the better the work is done the quicker it is done, because done systematically, and every stroke tells. No work lags so much as slovenly work.

One word more regarding large stones : When used, they must always be swung in place with the derrick, and *never* barred nor rolled to their bed over fresh work. This must be insisted on, or the bond of the mortar will be surely destroyed by jarring and displacing the stones last laid.

In dry riprapping, as in concrete, the best results attend the use of stones of various dimensions, so disposed that the percentage of voids shall be as small as possible.

When laying masonry in wet excavations see that the pumping sump is sunk below the bottom course of masonry, so that the pumps shall not draw the mortar out of the work, and placed outside of the area to be built over. This is a very important, though almost always neglected, point. In backfilling an excavation after the masonry has been built in it, see that the earth is well compacted, dead against the wall, so as to leave no vacancy between it and the sides of the excavation. This is also an important and generally neglected matter.

When work of this kind is undertaken it is indispensable that the principal engineer should give his close personal attention to all the details, particularly at the commencement, because both the contractors and the inspectors must be not only *told*, but *shown*, what is required to be done. Later on, when it may be impossible for him to give his undivided attention to the execution of the work, he should make frequent and prolonged visits to it, at irregular times. It is no reflection on the honesty or ability of the contractors and engineering staff to say that this is indispensable to secure good results; it is merely another way of saying that if the chief neglects his duty he cannot expect that the others will properly perform theirs.

One of the principal engineering difficulties which dam-building presents is what to do with the water while the work is in progress, and particularly while the foundations are being put in. When dealing with a powerful stream the problem assumes formidable proportions. Since it is indispensable that all dams should have a capacious blow-off culvert, at or near the bottom of the reservoir, I think the best way will often be to get this culvert and all its appurtenances in first, so that the stream may be diverted into it, and thus give no further trouble in average stages of the river. Should the season of freshets intervene when the culvert may not be able to handle all the water coming down to it, the work must be so prepared that the flood may sweep over it without damage. Much forethought is necessary in such cases, and more or less risk can hardly be avoided; the important point being to reduce it to a minimum.

Another difficulty occurs when springs are encountered in the foundation pits. These may be sometimes dealt with by providing a temporary escape for them through the masonry, to be closed afterward when the surrounding masonry has thoroughly set. They can also sometimes be overmastered by pumping, but the ingenuity of all hands is frequently taxed to devise means of getting rid of them, and the firmness of the engineer will also often be severely tested in refusing to allow the work to be commenced without adequate preparation, or to permit a defective foundation, through which a current of water is passing, to remain and receive its superstructure while in that condition.

The above comprise some of the most essential points to be observed in dam-building, but it is impossible to cover the whole ground, and the engineer intrusted with such work will find all his experience and resources called into constant requisition.

One governing principle should never be lost sight of in designing such structures. While thoroughly good work should be exacted in the smallest detail, yet the design of the dam should be such as to provide for disaster arising from overlooked defects of workmanship or from unforeseen contingencies, in such a way as to prevent or at least to minimize the consequent destruction. In the case, for instance, of a cloudburst gorging the spillway of an earthen dam, when the absence of the guardian has prevented the relieving blow-off culvert from being opened in time, there is almost always some point at which the breach can be made with the least bad results, and the design should favor its occurrence at this point rather than in less advantageous ones. I may add that the existence of a stout masonry center wall is always a tower of strength, and may frequently prevent a catastrophe should the dam be overtopped. "And even if the worst should "occur, and the center wall be breached, time, that priceless "element in such cases, would be gained, and the catastrophe "greatly modified. For the only difference between the harm-"less emptying of a reservoir, and a Johnstown disaster, is one "of time." *

---

* Discussion of Mr. Desmond Fitz Gerald's paper on "Rainfall, Flow of Streams, and Storage," Transactions of the American Society C. E., Vol. XXVII., page 295, which see also for further remarks about masonry center walls and capacities of spillways.

NOTES TO PARTS I. AND II.

Page 18.—It will be observed that (1) may be written:

$$V = \sqrt{\frac{D \times H}{C \times L}} \qquad \text{(1) bis.}$$

This shows that V increases with the square root of the diameter and of the head, and diminishes with the square root of the coefficient and length.

If diameter d be given in inches, then to obtain V in feet, multiply C by 12. Thus:

$$V = \sqrt{\frac{d \times H}{12\,C \times L}} \qquad \text{(1) ter.}$$

Page 30.—The general form of an adfected quadratic is:

$$x^2 \pm a\,x = b$$

Whence:

$$x = \mp \frac{a}{2} \pm \sqrt{b + \left(\frac{a}{2}\right)^2}.$$

Page 31.—Throughout this book proportions are written in the fractional form, as being the most convenient. The proportion $\frac{h}{22} = \frac{50}{54.87}$, is read: "$h$ is to 22 as 50 is to 54.87."

Page 42.—All the relations between the different elements of two long pipes may be expressed in a single equation:

$$\frac{D'^5 . H' . Q^2 . L . C}{D^5 . H . Q'^2 . L' . C'} = 1.$$

Assuming a mean common value for C and C':

$$\frac{D'^5 . H' . Q^2 . L}{D^5 . H . Q'^2 . L'} = 1 \qquad \text{(a)}$$

The most useful outcome of the above is when two pipes have the same length and head; that is, when $L = L'$ and $H = H'$. Then:

$$\frac{Q'}{Q} = \sqrt{\frac{D'^5}{D^5}}$$

That is, other things being equal, the quantities discharged by two pipes are as the square roots of the fifth powers of their diameters.

It is generally easier and more satisfactory to ascertain what we want to know about a given pipe by direct calculation, rather than to deduce it from the known elements of another pipe, by means of the relation (a).

Page 47.—The great practical importance of the formulæ on this page has not been brought into sufficient prominence in the original text. A more thorough elucidation will be given now.

Equation (3), page 18, can be written thus:

$$Q = \sqrt{\frac{0.616\, D^5.\, H}{C.\, L}}.$$

Consulting the table of coefficients on the same page, we see that those for rough pipes from 8 to 48 in. in diameter vary from 0.00068 to 0.00062. These coefficients do not greatly differ, and, moreover, their square roots only enter the formulæ. If, then, for pipes of the above diameters we take a uniform value, $C = 0.000616$, the above formula reduces to:

$$Q = \sqrt{\frac{1000\, D^5.\, H}{L}} \qquad (b)$$

Furthermore, since $\dfrac{H}{L}$ is a ratio, we may always reduce H to its value when $L = 1000$; that is to say, we can establish the relation:

$$\frac{h}{1000} = \frac{H}{L}$$

in which h equals the fall, or head, per 1000. Therefore, replacing $\dfrac{H}{L}$ in (b) by its equal $\dfrac{h}{1000}$, we have:

$$Q = \sqrt{D^5 h} \qquad (c)$$

Always recollect that h represents the head per 1000, as against H, which represents the total head in the total length.

The relation (c) may be generalized thus:

$$\frac{Q^2}{h D^5} = 1 \qquad (d)$$

Again, from table, page 18, we see that the value of C, for rough pipes from 3 to 6 in. in diameter, varies from 0.00080 to 0.00072. Also, equation (3), on same page, may be written:

$$Q = \sqrt{\frac{0.785 \times 0.785\, D^4 \cdot H}{C \cdot L}}$$

Adopting, therefore, a mean value of 0.000785 for C, for pipes of above diameters, and reducing $\dfrac{H}{L}$ to $\dfrac{h}{1000}$, as before, we have:

$$Q = \sqrt{0.785\, D^5 h} \qquad (e)$$

and generalizing,

$$\frac{Q^2}{h D^5} = 0.785 \qquad (f)$$

For smooth pipes, from 8 to 48 in. diameter, (d) becomes:

$$\frac{Q^2}{h D^5} = 2 \qquad (g)$$

For smooth pipes from 3 to 6 in. diameter, (f) becomes:

$$\frac{Q^2}{h D^5} = 1.57 \qquad (h)$$

From the above, we see that the discharge of a smooth pipe is 1.40 times that of a rough one of same diameter, while that of a rough one is 0.70 times that of a smooth one.

Observing that velocity is always equal to quantity discharged divided by area of pipe, we have for the velocity of flow, in feet per second, for rough pipes, from 8 to 48 in. diameter :

$$V = 1.27 \sqrt{Dh} \qquad (i)$$

For those from 3 to 6 in. diameter :

$$V = 1.13 \sqrt{Dh} \qquad (j)$$

For smooth pipes respectively :

$$V = 1.80 \sqrt{Dh} \qquad (k)$$

and

$$V = 1.60 \sqrt{Dh} \qquad (l)$$

The relation between the diameters of rough and smooth pipes, for equivalent discharges, may be also deduced, giving :

$$\frac{\text{Rough Diameter}}{\text{Smooth Diameter}} = 1.15 \qquad (m)$$

That is, the diameter of a rough pipe to give the same discharge as a smooth one should be 1.15 times that of the latter.

It is to be remarked that if $q = $ U. S. gallons per minute, and $d =$ diameter of pipe in inches, we have :

$$\frac{q^2}{h\,d^5} = 0.81 \qquad (n)$$

This is an intermediate value between (d) and (f), so that in a great many cases it would be quite safe to use (d) or (f) indifferently for feet and seconds, or for gallons, inches and minutes.

Comparison with observed velocity through pipes of different diameters and different degrees of roughness and smoothness* seems to establish the fact that the above series of formulæ, from (a) to (n), both inclusive, cover the whole range of practical cases, and that, apart from entirely abnormal conditions, no cast-iron pipe, however smooth, will give a greater discharge than that deducible from (g) and (h), nor will any or-

---

* See discussion of Mr. Desmond Fitzgerald's paper, "Flow of Water in 48-in. Pipe," Vol. XXXV., Transactions American Society Civil Engineers, July, 1896.

dinary degree of roughness reduce the discharge below that given by (d) and (f).

The above series of formulæ, therefore, may be confidently used in all practical hydraulic calculations, to the exclusion of all others given in the first part of this book, with no sacrifice of accuracy, and a great gain in simplicity.

It must be again repeated that since the tendency of all pipe lines is to become more or less incrusted with age, and as, besides, many other causes, such as leakage, etc., tend constantly to diminish the quantity discharged by any given pipe line, it is no more than common prudence to adopt always the formulæ for *rough* pipes.

Page 48.—As an example of the use of the table on this page, let it be required to know the diameter of pipe necessary to discharge 12 cu. ft. per second, with a fall of $\dfrac{3}{1000}$. From (d), we have:

$$D = \sqrt[5]{\dfrac{144}{3}} = \sqrt[5]{48}$$

Consulting the table, a diameter of 26 in. is found to correspond to the fifth root of 47.75. Twenty-six inches, therefore, is the proper diameter.

All that precedes has had reference to *long pipes*, when only the frictional head—erroneously so called—has been considered. In the case of *short pipes*, the exclusive consideration of this head is no longer admissible. The method to be then pursued will be best illustrated by an example.

Let it be required to find the total head necessary to discharge 96 cu. ft. per second through a 36-inch pipe, 150 ft. long. Calling the area of the pipe 7 sq. ft., the required velocity will be $\dfrac{96}{7} = 13.70$ ft. per second. The head necessary to produce this velocity, according to the laws of falling bodies, is $\dfrac{V^2}{2g}$. We have already seen (page 12) that the head necessary to overcome resistance to entrance is about one-half of this, so the two combined would be $\dfrac{3 V^2}{4g}$. The frictional head per 1000 from (i) is $\dfrac{V^2}{1.61 \times 3}$ which for a length of 150 ft. reduces to $\dfrac{15 V^2}{483}$. The velocity being 13.70, $V^2 = 187.69$. Therefore:

$$\text{Velocity and entrance head} = \frac{3 V^2}{4g} = \frac{3 \times 187.69}{4 \times 32.2} = 4.37 \text{ ft.}$$

$$\text{Frictional head} = \frac{15 \times 187.69}{483} = 5.83 \text{ ft.}$$

Total head above center of pipe 10.20 ft.

If the pipe did not discharge freely in air (see page 57) there would be back pressure equal to entrance head, and we would have:

$$\text{Velocity, entrance and exit head,} \ \frac{V^2}{g} = 5.83 \text{ ft.}$$

$$\text{Frictional head, as before } 5.83 \text{ ft.}$$

Total head, above center of pipe 11.66 ft.

As an additional example, how many cubic feet per second would a pipe, 24 in. diameter and 30 ft. long, discharge freely in the air from a reservoir, its center being 12 ft. below the surface of the water?

The velocity and entrance head, using round numbers, is $\dfrac{3 V^2}{4g} = \dfrac{V^2}{43}$. The frictional head per 1000 from (i) is $\dfrac{V^2}{3.22}$.

The friction head for the given length, 30 ft., will be $\dfrac{3 V^2}{322}$.

The total head above center of pipe being 12 ft., we have :

$$12 = \frac{V^2}{43} + \frac{3\,V^2}{322} = V^2 \left(\frac{1}{43} + \frac{3}{322}\right)$$

Using round numbers :

$$12 = V^2 \left(\frac{1}{43} + \frac{1}{107}\right) = V^2 \left(\frac{107 + 43}{4600}\right)$$
$$V = 19\text{ ft. per second.}$$

The quantity discharged being equal to the velocity multiplied by the area of the 2 ft. pipe is $19 \times 3.14 = 59.66$ cu. ft. per second.

It will be observed that "round numbers" have been freely used in the above calculation. A great deal of unnecessary figuring can be avoided, and an ample degree of accuracy secured, by a judicious discarding of small decimals in all hydraulic work.

Page 70.—Many other useful data can be added to the few already given. Thus:

Cu. ft. per sec. × 86400 = cu. ft. per 24 hours
" " × 646272 = U. S. gallons per 24 hours.
" " × 26928 = " " " hour.
" " × 448 8 = " " " minutes
$\left.\begin{array}{l}1.55 \text{ cu. ft. per sec.}\\93.00 \text{ " " min.}\end{array}\right\} = 1001722$ U. S. gallons per 24 hours.

It is therefore convenient to remember that 1.5 cu. ft. per second equals very closely 1,000,000 gallons per 24 hours.

Also :

1 acre covered 1 in. = 3630 cu. ft. = 27,152 U. S. gallons.
1 " " 1 ft. = 43560 " = 325,829 " "
1 square mile covered 1 in. = 2323200 cu. ft. = 17377536 U. S. gallons.
1 " " " 1 ft. = 27878400 " = 208530432 " "

Also :

$$HP = \frac{GH}{4000}$$

HP = net or theoretical horse-power.
G = U. S. gallons per minute.
H = height in feet to which water is raised.

Also, for discharge over weirs or spillways :

$Q = 2000000\,L\,\sqrt{H^3}$.
Q = U. S. gallons per 24 hours (close approximation).
L = length of weir in feet.
H = height from sill of weir to surface of still water, in feet.

The horse-power, HP, of water falling over a weir:

$$HP = 0.35\, L \times F\, \sqrt{H^3}.$$
$F =$ fall in feet, other symbols as above.

Also, for weight in long tons of cast-iron pipe, per mile, we have:

$P = 0.20\, w\, M.$
$P = 25\, M\, (D + T)\, T.$
$P =$ approximate weight in long tons (2,240 lbs.) of pipe line.
$w =$ weight in lbs. of 12 ft. length over all of pipe.
$M =$ length of pipe line in miles.
$D =$ inside diameter of pipe in inches.
$T =$ thickness of pipe in inches.

All the formulæ given in this book refer to cast-iron pipe. We are as yet without sufficient experimental data as to how far they may be applicable to wrought-iron or steel-riveted pipe. It appears, however, that such pipe do not give as high velocities as cast-iron pipe of the same diameter. It is probable that the diameters of wrought-iron or steel lap-jointed riveted pipe should be from 5 per cent. to 10 per cent. greater than those of cast-iron pipe, to insure an equal discharge.

Page 71.—Chemical science has not yet reached the point of drawing definite conclusions from an analysis of water. Still a chemical examination should always form part of the investigation of the quality of any proposed water supply, as it is frequently of great utility, when combined with other information, in enabling an intelligent judgment to be formed of the wholesomeness of the supply.

The element of danger most to be dreaded in a water supply is sewage contamination, and it is to the detection of the evidences of such contamination that the efforts of the chemist are mainly directed in an examination of a water sample. The substances regarded with the greatest suspicion are albuminoid ammonia, chlorine, and nitrites. In regard to these, the following quotations from Mr. C. C. Vermeule's Report, forming Vol. III. of Final Report of the State Geologist of New Jersey, will be found very useful:

"*Albuminoid Ammonia.*—This represents animal and vegetable matter present in the waters and in process of decomposition, by which process free ammonia is produced, consequently it is more to be dreaded than the latter, as at certain stages of decomposition such matter becomes very dangerous to health. Dr. Leeds' limit is 0.028 (parts in 100,000).

"*Chlorine.*—This is not only an accompaniment of sewage pollution, but is a measure of the amount of such pollution, although not always of the danger to be apprehended therefrom. Dr. Leeds gives the maximum allowable at one part in 100,000.

"*Nitrates and Nitrites.*—Pollution by sewage being practically the addition of nitrogen compounds to the water, the process of purification of this water consists of the oxidization of these compounds, and when this process is completed they become nitrates. Nitrites indicate that this work of purification is in progress, but is not complete, consequently their presence is a more serious matter than that of nitrates."

Mr. J. T. Fanning, in his valuable work on water-supply engineering, gives as a quotation Heisch's "Simple Sugar Test of Water," as follows:

"If half a pint of the water be placed in a clean, colorless-glass stoppered bottle, a few grains of the best white lump sugar be added, and the bottle freely exposed to the daylight in the window of a warm room, the liquid should not become turbid, even after exposure for a week or ten days. If the water becomes turbid, it is open to grave suspicion of sewage contamination; but if it remain clear, it is almost certainly safe."

Page 74.—Within the last few years, indeed since the appearance of the first edition of this little volume, two features of water-supply engineering hitherto but little considered in this country have forced themselves into notice. These are artesian, or driven wells, and the filtration of public water supplies.

 Artesian wells have, it is true, been long used for small supplies, but of late years their use on a large scale has taken a great development. In many places they afford the only means of supply, and after a vain attempt to secure surface water in sufficient quantities and of a proper quality, it becomes frequently necessary to have recourse to the vast supplies stored away beneath the surface, and which may be exploited or mined like any other subterranean deposit of valuable material.

The employment of these wells has natural limitations. They require the existence of a suitable geological formation, without which they would be unproductive. They must reach a permeable, deep-seated stratification, which is found only, and by no means always, in the secondary and tertiary formations. Of these, the supply taken from the secondary is, while of greater rarity, generally of more considerable volume than that furnished by the tertiary.

In some of these artesian wells the water reaches to the surface of the ground, and even spouts out to a considerable height above it. This greatly facilitates the practical utilization of the supply. In others the water reaches only to a considerable depth below the surface, when the difficulties of raising and distributing it become greatly increased. If the whole supply is furnished by a single boring the problem is much simplified, because the pump can be placed directly over the well and a lifting main used to bring the water within reach of the force main, without the employment of a separate pump.

If, however, as is generally the case, a gang of wells is necessary, the expense of getting the water to the force main is greatly increased, unless the water in the wells rises to within, say, 20 ft. of the surface. When this occurs, the yield of the whole gang can be collected in a single suction main, feeding to the pump. Otherwise each well must have a separate pump. It is probable that in such cases power might be advantageously transmitted electrically from a single engine to all the auxiliary pumps, or some system of air lifts might be used, operated by a single compressor.

In studying the project of a driven-well system for any particular locality, although general inferences may be drawn from the geology and topography of the district, no positive knowledge can be obtained without sinking experimental wells. Without such practical tests, all forecasts are but little better than guesswork.

Page 76.—In estimating the amount of water required for any locality, the quantity needed for fire service must be carefully considered. For small towns, it will be found that if provision is made, both as to quantity required per minute, and pipe capacity to convey the same, in regard to fire service, it will more than cover all other needs, while in the case of a large city the water needed for fire service is a relatively small proportion of the whole, and if all the requirements of domestic and the other public needs are met, no further provision is necessary for fire service.

According to Mr. Fanning, a 4-in. pipe is required to supply a single hose stream of about 20 cu. ft. per minute. A 6-in. pipe, according to the same authority, will furnish two such streams. It is quite clear, therefore, that for the smallest town a 4-in. main is the absolute minimum, while for towns of any considerable size, either actual or prospective, 6 ins. is the smallest sized pipe that should be laid in any of its streets. As a general rule, it may be stated that each street main should be sufficient to furnish enough water for at least one ordinary conflagration in that street.

Page 77.—All extensive exposed water surfaces, such as lakes and large ponds, lying within the survey, should be excluded from the estimate of available watershed area, because the evaporation from such surfaces is practically equal to the rainfall upon them.

Page 80.—A convenient formula, and one which it is believed will prove in most cases to be approximately correct, for the necessary amount of storage is the following:

$$S = \frac{C^2}{Y}$$

In this formula, $S =$ the amount of storage required, $C =$ the yearly consumption, and $Y =$ the yearly available yield of the watershed, all expressed in the same unit.

Example.—A town consumes 10 million gallons per day, or 3,650 millions per year. The estimated yearly yield of the source whence the supply is drawn is ten times that amount, or 36,500 millions. What storage capacity is needed to insure the required daily amount throughout the year?

$$S = \frac{13322500}{36500} = 365 \text{ million gallons.}$$

This is equivalent to storage of 36.5 days' supply, or that of about five weeks.

Page 83.—In the calculations given on this and subsequent pages, no account has been taken of the resistance to sliding upon the base. It is not generally necessary to do so in the case of masonry dams, because such structures always are, or should be, built upon a rock foundation, and therefore cannot be pushed forward without shearing through the joints of the masonry. If a wall rests on a slippery foundation, however, there may be great danger of its sliding forward long before it can yield by overturning. The resistance which a wall or dam sustaining water pressure offers to sliding is its weight multiplied by some "coefficient of friction," always taken as less than unity, while the force tending to produce sliding is the thrust of the water $= 31.25$ H^2, H being the depth of water pressing against the wall.

If we assume a density of 125 lbs. per cubic foot as that of the wall, and represent its thickness by B, assuming also a coefficient of friction of 0.75, then if a plumb wall sustains water pressure against its full height H, the equation of equilibrium is:

$$31.25\, H^2 = 0.75 \times 125 \times H \times B.$$

$$B = \frac{H}{3}$$

A thickness equal to two-thirds the height of such a wall would therefore give a factor of safety of 2, as against sliding, but this factor would be reduced if the wall were transformed to a trapezoidal section according to Vauban's rule, because the weight would be diminished.

Two excellent formulæ, readily deducible from the principles already laid down, may be used for determining the bottom width, b, of a trapezoidal wall sustaining water pressure to its full heighth, h, when its top width, t, and the density, d, of the material of which it is composed are given, as well as the desired factor of safety, f, and the coefficient of friction, c. They are, as against sliding :

$$b = \frac{62.50\, h f}{c\, d} - t$$

and as against overturning :

$$b = \frac{1}{2}\sqrt{\frac{125\, f\, h^2}{d} + 3\, t^2} - \frac{t}{2}$$

Example.—Let $h = 30$ ft., $t = 6$ ft., $d = 125$ lbs., $f = 2$, and $c = 0.75$. Then, as against sliding :

$$b = \frac{62.50 \times 30 \times 2}{0.75 \times 125} - 6 = 34 \text{ ft.}$$

As against overturning :

$$b = \tfrac{1}{2}\sqrt{\frac{125 \times 2 \times 900}{125} + 3 \times 36} - \frac{6}{2} = 18.85 \text{ ft.}$$

These results bring out in bold relief what has been already said regarding possible danger of sliding when the wall is not well tied down to a solid foundation.

WATER SUPPLY ENGINEERING. 119

A few memoranda regarding hydrostatic pressure against wet surfaces will be useful in this connection.

As regards pressures against plane vertical surfaces, nothing need be added to what has already been said in reference to dams (page 83, etc.), but some additional information regarding inclined and curved surfaces will be useful.

Pressure upon the plane inclined surface, of which the edge is shown by the line AB (fig. 22), and of which the length, at

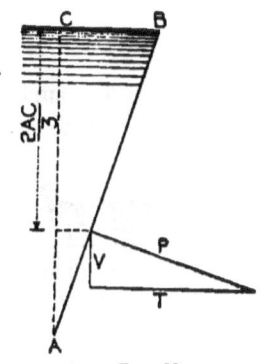

FIG. 22.

right angles to the plane of the paper, is supposed for convenience to be 1 foot, is of three kinds. First we have *total pressure* in pounds which is equal to the length of the line AB multiplied by half the height AC, and by 62.5. The point of application of this pressure is at the distance $\dfrac{AB}{3}$ from A, and is exercised in a direction normal to AB.

This total pressure is divided at the point of application into two components, which form the two other pressures referred to, one vertical and downward, and the other horizontal and outward. The first of these is the total vertical downward pressure upon the surface represented by AB, and is equal to the weight of water resting upon it, or the area of triangle ABC, multiplied by 62.5 lbs., and the other is the horizontal thrust

and is equal to $\dfrac{\overline{AC}^2}{2}$ multiplied by 62.5 lbs. $= 31.25\,\overline{AC}^2$. The

overturning moment of this latter pressure is $\dfrac{62.5\,\overline{AC}^3}{6} =$ $10.42\,\overline{AC}^3$ ft. lbs.

The pressures upon the curved surface AB (Fig. 23) are naturally more complicated. The total pressure is best deter-

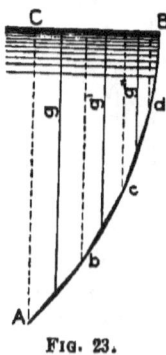

FIG. 23.

mined graphically by dividing AB into a certain number of short, equal straight lines Ab, bc, etc., thus converting the curve into a nearly equivalent polygon. The total pressure upon these elements will be $Ab \times g$; $bc \times g'$, etc., g, g', g'', etc., being the vertical distances from the middle of the lines Ab, bc, cd, etc., to the surface of the water, and the total pressure upon AB will be given by their sum. The total vertical and horizontal pressures will be the sums of those of the elements Ab, bc, etc., and will be, therefore, the one equal to the weight of the volume of water $A\,B\,C$, and the other equal to $\dfrac{\overline{A\,C}^2}{2}$ the same as for the plane inclined surface.

The point of application of this horizontal thrust is at the height $\dfrac{A\,C}{3}$ from A. It will be observed that both the thrust and its point of application are the same for vertical and inclined plane surfaces, and also for curved surfaces.

The point of application of the vertical downward pressure, both for plane inclined and curved surfaces, passes through the centre of gravity of the volume of water resting upon the surface.

Page 87.—In the author's opinion the two formulæ (9) and (10) may be replaced by the single general one :

$$P = \left(\dfrac{L-D}{L\,.\,D}\right)W.$$

This formula agrees with both of the others for $D = \dfrac{L}{3}$, and with (10) for $D = \dfrac{L}{2}$. It also agrees very closely with (10) for all intermediate values of D. It gives values increasingly greater than (9) for all values of D between $\dfrac{L}{3}$ and o, which in the author's opinion it should do, since we are warned by good authorities that much confidence is not to be placed in (9) when D becomes more than slightly less than $\dfrac{L}{3}$. The author is also inclined to the belief that the pressure is distributed more or less uniformly throughout the entire length of D, and not concentrated at its extremity.

Page 92.—In regard to the section shown in Fig. 19, it would be improved by making the slope DB a little steeper, and the slope CE a little flatter, than shown, so as to increase the unit stress at D and reduce it at E. It is safer, also, in these calculations to discard the moment of the water pressing vertically downward against the inclined inner side of the dam, as its action through the mass of masonry is somewhat uncertain. It will always be an additional factor of safety, but it is doubtful if its amount can be accurately determined as to its practical action.

Pages 98–99.—In regard to the manner of drawing water from a reservoir, sliding sluice gates have been of late so perfected in their design, and have proved so efficient and reliable in practice, that one of the valves recommended in the text may be advantageously replaced by a sluice gate of approved make, set against the mouth of the pipe, inside of the tower.

THIRD PART.

FLOW OF WATER THROUGH MASONRY CONDUITS, AND SOME DETAILS OF TUNNEL CONSTRUCTION.

From his experimental researches, Darcy deduced formulæ for the mean velocity of water flowing through open canals, or any other form of conduit laid to a uniform grade, and not running under pressure. The formulæ vary with the degree of smoothness of the interior surface of the conduits. The formula which he established for smooth, though not polished, surfaces, and which is the one suited most nearly to well-constructed, brick-lined aqueducts, is as follows:

$$\frac{R I}{U^2} = 0.00019 \left(1 + \frac{0.07}{R}\right)$$

In which R = hydraulic mean radius, or the water section divided by the "wet perimeter"; I = the slope per unit of length, or the total fall divided by the total length, and U the mean velocity of the current. This formula applies to the metric system, and U is expressed in meters per second.

To adapt the above to feet, it may be most conveniently written as follows:

$$\frac{R I}{U^2} = 0.00001 \left(6.6 + \frac{0.46}{R}\right) \quad (1)$$

Whence

$$U = R \sqrt{\frac{100000\, I}{6.6\, R + 0.46}} \quad (2)$$

In which U = mean velocity, in feet per second.

As a simple illustration of the use of this formula, suppose a canal, 6 ft. wide, with vertical sides, laid to a slope of $\frac{1}{1000} = 0.001$ and running 3 ft. deep. The water section is then 18 sq. ft., the wet perimeter 12 ft., and the mean hydraulic radius is 1.5.

Substituting the above data in (2):

$$U = 1.5 \sqrt{\frac{100000 \times 0.001}{6.6 \times 1.5 + 0.46}} = 4.67 \text{ ft. per sec.}$$

It is to be noted that the above formulæ (1) and (2) apply only to conduits of sufficient length and uniformity to permit of the establishment of what is called the "permanent regimen." This occurs for any canal or channel, when the depth of water is uniform throughout its entire length; that is, when the surface of the water is parallel to the bottom of the conduit.

In the special case of a circular section, running full but not under pressure, $R = \frac{D}{4}$, D being the diameter in feet. It will be interesting to see how the velocity through a brick-lined conduit 4 ft. in diameter, as given by the above formula, compares with that of a smooth cast-iron pipe of the same diameter. Assuming a fall of $\frac{1}{1000}$ in both cases, we have for the brick conduit:

$$U = \sqrt{\frac{100000 \times 0.001}{6.6 + 0.46}} = 3.76 \text{ ft. per sec.}$$

For a smooth cast-iron pipe of same diameter, we have from x (k):

$$V = 1.80 \sqrt[4]{4} = 3.60 \text{ ft. per second}$$

This is slightly less than that through the brick-lined conduit. The increased velocity through such conduits seems to be due to the fact that they are laid to a true uniform grade, with few changes of direction, and present a uniform and unbroken interior surface.

For conduits lined with good rubble masonry, the mean velocity U' may be taken as:

$$U' = U \sqrt{\frac{19R + 1.33}{24R + 60}}.$$

U being the mean velocity through a brick-lined conduit of the same elements. U' will generally be from 80% to 90% of U.

The application of a formula to such a conduit is very uncertain, because the projections of the rubble masonry render it impossible to determine the exact area of waterway.

The maximum discharge through a circular or a "horseshoe" shaped conduit does not occur when running entirely full (unless it is under pressure), but when full to within about $\frac{1}{10}$th of the radius of the crown, because the velocity and the wet section are then in the most favorable relation to each other. Thus, the discharge through a circular conduit 6 ft. in diameter would be greatest when it was running full to within 0.30 ft. of the top. In a circular conduit the velocity, when running half full, is precisely equal to that when running entirely full, so the discharge is just one-half. It is said that the most favorable section of the horseshoe type is when the total height is equal to the greatest width, which will naturally be at the springline of the arch. This assimilates the section to a circle.

As to the best form of cross-section for masonry conduits of large dimensions, the circle has much to recommend it as regards strength, economy, and delivery. It can be readily lined with bricks of ordinary shape and dimensions. On the other hand, the horseshoe form affords more floor room, and possesses certain facilities for construction, among others that of permitting the building of the sides and arch of the lining before the invert is put in, as was very successfully done in building the new Croton aqueduct. The driving of a tunnel, when the excavation and masonry lining are carried on simultaneously, is an exceedingly troublesome piece of work, on account of the confined space in which the operations are executed and the great amount of materials passing in and out. If the invert is built first, as would seem to be the natural order of executing the work, it is exposed to injury, and greatly impedes drainage and also the hauling of materials.

For these and other reasons it is a great advantage to leave the invert till all the rest of the work is completed from shaft to shaft. The horseshoe form permits of this very well, but the use of special bricks is required to make the junction between the side walls and the invert. Such special bricks were used in the new Croton aqueduct. A course of cut stone blocks or skew-backs would be far better than the special bricks, but very much more expensive.

One of the most important details of tunnel construction is the backing behind the side walls and over the arch. Nothing is really satisfactory but wet masonry of some sort, although for economy dry work is sometimes, perhaps generally, specified. The effect of dry stone packing over an arch is really little more than loading it very irregularly with a heap of loose stones and small material. At all events the spandril backing should be carried up to the level of the extrados of the crown of the arch in cement masonry. All small vacancies are probably best taken up with wooden lagging packed tight. In a tunnel this lagging lasts a long time and affords an elastic cushion to deaden the shock of any movement of the roof which may occur. By the time it decays, if it ever does decay, the roof will have adapted itself to its final bearings, and the cement of the masonry lining and walls will have become thoroughly indurated.

The alignment of the new Croton aqueduct demonstrated the surprising accuracy with which the centre line of a tunnel can be ranged between plumb lines dropped from the two sides of a shaft. As a useful detail, it may be interesting to state that it was found that, owing to its optical properties, a telescope, placed in line with two plumb lines, with the vertical cross hair exactly covering both, could, if placed sufficiently near the nearest one, focus it entirely out of sight, the visual rays passing around it, on both sides, and focusing on the farther line. This property was exceedingly useful in setting the instrument in line.

Sewers are often built with an egg-shaped cross-section, the point downward. This form seems best suited to accommodate a variable flow, although the circle has much to recommend it for sewers also.

FILTRATION.

The development of the filtration of public water supplies is a recent feature of hydraulic engineering in this country. Hitherto this process has been looked upon with general disfavor and, in the case of new water-works, the endeavor has always been to secure an unobjectionable supply of natural water. This is perfectly proper, when practicable, but year by year the attempt becomes more difficult, and besides, supplies which when first used were perfectly wholesome may become so polluted by gradual encroachments as to be rendered unfit for use. In this manner attention is now being turned, in the United States, to the practical application of those processes of purification which have been so long and, as it seems, successfully, employed in Europe.

Another reason for the growing favor with which filtration on the large scale is now being regarded is the fact, recently established, that, contrary to previous belief, filtration is not confined in its action to a mere retention of matters held in suspension, but also exercises a marked chemical effect upon impure waters.

Although our knowledge of the true action of filtration has greatly advanced, it is still far from being complete, and the precise process by which the chemical purification is effected is still under discussion. It seems however certain that the *sedi-*

mentation layer, or the gelatinous film formed by deposition from the water itself, plays a great part in that destruction of a large percentage of noxious bacteria which is found to be the result of a proper system of filtration.

Owing to the great areas required for filter beds—about one acre per day, per $2\frac{1}{2}$ million gallons, "mechanical filters" (so-called) are largely used in the United States. These filters are much more rapid in their action than the ordinary filter beds, with the necessary result of being less efficacious. In order to increase their efficiency, a coagulant—generally alum—is often used. The sulphate of alumina precipitates rapidly many of the impurities of water, and then disappears with them wholly or nearly from the effluent.

The following particulars are given by the manufacturers of a leading mechanical filter : A fixed solution of sulphate of alumina is used, one pound sufficing for the clarification of from 50,000 to 150,000 lbs. of water. The filtering material is crushed crystal white quartz, "assisted, after the introduction of the sulphate of alumina, by the gelatinous mineral film of hydroxide of alumina. This sensitive membrane, with its fine quartz foundation, is so powerful as to be capable of retaining all coloring matter and bacteria." For city supply, the unit filter passes ordinarily 250 gallons per minute. When clean, one foot of head passes this quantity. As the filter silts up the necessary head increases progressively. When it reaches 8 ft. cleaning becomes necessary. Cleaning is effected by passing a sufficient quantity of water through the crushed quartz while it is being thoroughly stirred up by a special appliance.

"Initial cost of plant exclusive of land, foundations and buildings, depends on velocity of filtration ; it will average, delivered, erected and connected, $6,000 per million gallons. When the plant is cared for by an engineer required for other purposes as well, the cost of maintenance is about $2.50 per million gallons."

On the subject of filtration consult Hazen's "Filtration of Public Water-Supplies" and Mason's "Water-Supply."

Pumps and Pumping Engines.

In many instances the source of a water supply lies lower than the locality to which it is to be delivered. In such cases pumping must be resorted to.

A pumping plant consists essentially of a suction and force main, with a pump working between them.

It is clear that the suction valves of the pumps must always be sufficiently close to the level of the supply to ensure a steady and powerful draft at the suction end. This necessitates the placing of the whole plant at a low level, and when the elevation of the water in the sump is subject to fluctuations, as in the case of its being fed from a river, which may rise and fall considerably according to the season, the action of the pumps may be seriously impeded. The old Cornish type of pumping engine is comparatively free from this difficulty, because the plant can be located at any distance above the suction chamber, but for many reasons this type of engine is not now employed in this country for pumping large water supplies.

Pumping engines may be divided into the three classes of low, medium and high duty. This classification refers entirely to the relative consumption of fuel to accomplish a given delivery of water. Broadly speaking, a low duty engine is one consuming more than 4 lbs. coal ; a medium duty engine, one that consumes between 2 and 4 lbs., and a high duty engine, one that works with 2 lbs. or less of coal, per hour per horse power. It is evident that the low duty type includes all single cylinder engines, that the medium duty type calls for a compound engine, while the third or high duty class demands the most refined type of triple expansion engine using an automatic cut-off, and all other heat saving appliances.

The selection of one of these types in any particular case requires a great deal of judgment, and depends upon the required daily supply, the cost of fuel, facilities for repairs, the nature of the help, as well as the first cost of the plant. A high duty pumping engine is not only very expensive as to first cost, but it is, of necessity, a complicated piece of mechanism, requiring the best skill for its care and operation, and exceptional facilities for repairs in case of a breakdown. It must be borne in mind, too, that, notwithstanding all assertions to the contrary, the great economy of fuel of the modern high duty pumping engine is only realized in practice when it is working steadily at its maximum capacity. A varying supply is fatal to its best results.

In a word, the high duty pumping engine finds its most fitting field in the supplies of large cities, where great and nearly constant quantities of water are consumed daily, and where the facilities for repairs and the procuring of skilled labor are most abundant. The scale descends down to the small town or village, where the crudest and simplest type of engine suits best the case.

The actual duty of a pumping engine is ascertained by a test. This test consists in running the engine and pumps for a certain number of hours, and during that period measuring, weighing and timing everything that is done by, in and about the entire plant, and then interpreting the data thus collected.

The duty of a pumping engine is best expressed by the number of foot pounds of work actually performed per million heat units (B. T. U.) delivered by the boiler to the engine. Sometimes, though in this country less frequently now than formerly, duty is expressed in foot pounds of work per hundred pounds of coal consumed. The objection to this method is that it tests coal, boiler and engine as a whole, whereas their performances should properly be kept distinct.

The principal difficulty in these tests is in measuring the volume of water lifted. When the quantities are comparatively small they may sometimes be measured directly. Weir measurements come next in order of merit. Frequently the only available method is by plunger displacement, allowance being made for leakage and slip. As many checks as possible should be used in determining this important factor.

The volume, or weight of water lifted being ascertained, the next thing is the height to which it has been raised, or rather the pressure used to force it to this height. Clearly much more power would be required to force one million gallons 100 ft. high in a given time, through a pipe 8 in. in diameter than through one of 16 in. It is not, therefore, the height that is required but the actual pressure overcome by the engine. To ascertain this two gauges are used, a pressure gauge on the force main and a vacuum gauge on the suction, unless this latter is under a head, when a pressure gauge is applied here also. These pressures, multiplied by the total travel of the piston, give the foot pounds developed during the trial.

The total heat units supplied by the boiler are calculated by taking the weight of all the water fed to the boiler from all sources, and the difference in degrees Fahrenheit between the initial temperature of each several weight of water and the total heat of the dry steam delivered at boiler pressure to the cylinders.

The formula for duty is then:

$$\text{Duty} = \frac{W \times 1000000}{H}$$

In this equation $W =$ total work in foot pounds, and $H =$ total heat in British thermal units:

When the duty is estimated in work done in foot pounds per 100 lbs. of coal consumed, the formula is:

$$\text{Duty} = \frac{100 W}{C}.$$

Where $C =$ pounds of coal consumed during the time that

the work, W, is being performed. To convert the above into pounds of coal, P, per hour per horse power :

$$P = \frac{198}{\text{Duty in mil. ft.-pds.}}$$

Thus for an engine having a duty of 100 million foot pounds per 100 lbs. coal :

$$P = \frac{198}{100} = 1.98,$$

or practically 2 lbs. coal per hour per horse power.

Rating 1 lb. average coal as equal to the evaporation of 10 lbs. water or the development of 10,000 B. T. U., we may say that "high duty" implies not less than :

1,000,000 ft. pds. per lb. coal
100,000 " " water.
100 " " B. T. U.

Duty trials are perhaps most generally made in this country under actual working conditions, that is, the main feed is pumped from the hot well, and the jacket and separator water fed back to the boiler. This is more satisfactory as representing normal conditions, but complicates the measurements.

The management of a duty trial is a very intricate affair, and cannot be fully described here. The report of the committee on standard method of conducting duty trials of pumping engines, of the American Society of Mechanical Engineers, in its revised form, should be consulted in this connection.

As illustrating the general outline of a duty trial, the following example, condensed and simplified from one given in the above report, will now be instanced : A high duty compound pumping engine is supplied with steam at 135 lbs. absolute pressure. This corresponds to a total heat, above zero Fahrenheit, of 1,220.70 B. T. U. There is a separator on the main steam pipe. After passing through this separator the steam is found to still contain 1½ per cent. moisture. This moisture affects the latent heat of the steam (which at above pressure is 866.60 B. T. U.), so that its total heat above zero is :

$$1220.70 - 866.60 \times 0.015 = 1207.7.$$

Both cylinders are jacketed, and there is a reheater supplied with boiler steam. Water from jackets, separator, and reheater feed back to boiler. The different supplies of water fed to boiler during trial (10 hours), with their temperatures, are as follows :

Main feed, at 215°.. 18,863 lbs.
Low pressure jacket, at 225°................................ 615 "
High   "   "   and reheater, at 290°............ 815 "
Separator, at 340°.. 210 "

Total feed.. 20,503 lbs.

The total heat furnished by the boiler is therefore :

Main feed (1,207.7 − 215) 18,863 =........................... 18,725,300
Low pressure jacket (1,207.7 − 225) 615 =................... 604,361
High pressure jacket and reheater (1,207.7 − 290) 815 =..... 747,926
Separator, neglected.

Total B. T. U... 20,077,587

The net area of pump plunger is 308 sq. in., and the average stroke 3 ft.

Number of single strokes during trial.......... 24,000
Pressure by gauge on force main..................... 95.00 lbs.
"   vacuum gauge on suction main.............. 3.69 "
"   equivalent to difference of level of gauges.. 4.31 "

Total pressure........... .... 103.00 lbs.

The work done by the pump is therefore $308 \times 103 \times 3 \times 24000 = 2{,}284{,}128{,}000$ ft. lbs.

$$\text{Duty} = \frac{2284128000}{20077587} = 113765066.$$

Indicated horse power, as determined during trial, 128.15. Pump horse power, as above :

$$\frac{2284128000}{10 \times 60 \times 33000} = 115.36.$$

Hence :

$$\text{Efficiency} = \frac{115.36}{128.15} = 90 \text{ per cent.}$$

This is efficiency of the engine as regards work done by the pistons, not as regards heat utilized. Upon this latter basis, the calculation would be as follows: Since 1 B. T. U. = 772 ft. lbs. and one horse power = 1,980,000 ft. lbs. per hour, the number of B. T. U. per hour per horse power is $\dfrac{1980000}{772}$ = 2565. But the total B. T. U. furnished by the boiler, per hour, is 2007758.7. Hence, the theoretical horse power corresponding to the heat unit supplied is:

$$\text{Theoretical H. P.} = \dfrac{2007758.7}{2565} = 782$$

Therefore:

$$\text{Efficiency} = \dfrac{115.36}{782} = 0.1475$$

The above theoretical horse power is, of course, impossible of realization, for it supposes the temperature of the hot well to be reduced to zero. Suppose, in the above case, for round numbers, that the temperature of the hot well were 100°; the temperature of the steam, at 135 lbs. absolute pressure, is 350° above zero Fah. and 810° above zero absolute. Then, by Carnot's law, the efficiency of a perfect heat engine working between the given limits is:

$$\dfrac{350 - 100}{810} = 0.3086$$

Then, according to one view of the subject:

$$\text{Efficiency} = \dfrac{1475}{3086} = 0.48 \text{ nearly.}$$

According to another view, which considers only the latent heat of the steam as that theoretically utilizable, we have for the total theoretically possible work in foot pounds per pound of steam (or feed water), the following formula:

$$W = \dfrac{L H (T - t) 772}{T + 460}$$

In which :
- W = total possible work in foot pounds per pound of feed water.
- $L\,H$ = latent heat of steam at given pressure.
- T = temperature of steam at given pressure.
- t = temperature hot well or condenser.
- $T + 46\rangle$ = absolute temperature of steam.
- 772 = Joule's equivalent.

In the example :
$$W = 866.6 \times \frac{250}{810} \times 772 = 206510.$$

As there were 20,503 lbs. of steam furnished, or water fed :
$$206510 \times 20503 = 4234074530 \text{ ft. lbs.}$$

Comparing this with the work actually done by the pumps, gives :
$$\text{Efficiency} = \frac{2284128000}{4234074530} = 0.54 \text{ nearly.}$$

The usual English method of conducting a duty trial seems to be to feed the boiler exclusively from a measured tank, wasting the jacket and injection water. The quantity and temperature of the wasted discharge of the air-pump is also measured, and the quantity of heat necessary to raise it from its initial temperature to that of the hot well calculated and recorded as "rejected heat." This quantity of heat, reduced to B. T. U. per minute per indicated horse power, is known as " Donkin's Coefficient." When, however, the feed, injection, and jacket water are measured separately, a more accurate estimate is possible than when the total discharge of air-pump is used. To this rejected heat is added the heat utilized per minute per indicated horse power, which is
$$\frac{2565}{60} = 42.75$$
multiplied by the indicated horse power developed during the trial.

The rejected and utilized heat added together should equal the total heat reckoned from boiler consumption. It always falls short, and the balance is put down to errors and radiation.

As an example of the English method the following is given based upon a trial made by Professor Unwin. The wasted jacket water was measured separately, and a more accurate basis of calculation thus established than by the use of Donkin's coefficient. All calculations are made in this example for the total duration of trial and not reduced to "per minute."

Duration of trial	24 hours.
Total water pumped	200,000,000 lbs.
Lift, including friction	50 ft.
Work done	10,000,000,000 ft. pds.
Feed at 51°	108,500 lbs.
Jacket water wasted	16,900 lbs.
Feed used in work (108,500−16,900)	91,600 "
Injection water, at 50°	3,633,000 "
Condensed steam	91,600 "
Coal burned	11,000 "
Temperature of hot well	75°
" steam, 75 lbs. absolute	307°
Total heat of above steam	1208°
Indicated horse power	255.50
Pump horse-power	210.44

$$\text{Efficiency} \frac{210.44}{255.50} = 82.3\%$$

Rejected heat :

Injection, 3,633,000 (75−50)	90,825,000
Feed, 91,600 (75−51)	2,198,400
Jacket water, 16,900 (307−51)	4,326,400
Total rejected heat	97,349,800
Heat units utilized during trial, 255.5 × 2,565 × 24	15,728,580
Total heat accounted for	113,078,380
Total heat furnished by boilers, 108,500 (1,208−51) =	125,534,500
Heat accounted for =	113,078,380
Deficit (about 10%)	12,456,120

The duty, calculated per million heat units furnished to cylinders (108500 − 16900) (1208 − 51) is :

$$\text{Duty} = \frac{10000000000 \times 1000000}{91600 \times 1157} = 94356358 \text{ ft. lbs.}$$

The English practice is to calculate duty per cwt., or 112 lbs. of coal. Then :

$$\text{Duty} = \frac{10000000000 \times 112}{11000} = 101818102 \text{ ft. lbs.}$$

Apart from minor details, and considering only those of large capacity, there are two prominent types of pumping engines in use in the United States; namely, the direct acting, with or without a high duty attachment, and the rotative, or crank and flywheel engine. The merits of these two types are fully set forth in the descriptive pamphlets of their respective makers ; here it will suffice to briefly enumerate their respective claims, as follows:

The advocates of the direct-acting type claim a large reduction of weight by replacing (when high duty is required) the flywheel and crank shaft by a comparatively light special attachment, and a greater security against damage in case of an accident suddenly relieving the engine of its load from the fact that there is no dangerous momentum stored in heavy moving parts. Also, that high duty is realized through a greater range of developed power, that is, nearly the same economy is claimed when working at reduced, as at full speed.

The advocates of the rotative type claim that "its principal advantages are positive action of steam valves and cut-offs, and absolute full stroke of steam pistons and plungers under varying pressures of steam and water. In these engines, therefore, there can be no increase in the clearance spaces between the steam pistons and cylinder heads causing waste of steam, nor loss of capacity by deficient plunger displacement."

It is but simple justice to say that many magnificent specimens of both types are to be found doing excellent service all over the country.

Arches and Abutments.

Some of the grandest and most interesting examples of hydraulic engineering are to be found in the arched aqueducts of ancient and modern times. A study of the principles of the arch is therefore an essential part of the equipment of the water-works engineer.

The span being given, the starting point of all arch calculation is thickness or depth of the crown at the key. This dimension is fixed in practice by some one of the various empirical formulæ in general use. Although these have been deduced from existing structures, and the most approved ones can be supported by many examples, they exhibit considerable variation in their results. Here are five, with the names of the authorities who give them:

$$\text{Perronnet.} \quad D = 1 + 0.035\, S \qquad (1)$$
$$\text{Croizette-Desnoyers.} \quad D = 0.50 + 0.38\, \sqrt{R} \qquad (2)$$
$$\text{"} \quad \text{"} \quad D = 0.50 + 0.33\, \sqrt{R} \qquad (3)$$
$$\text{Boix.} \quad D = 0.75\, \sqrt[3]{S} \qquad (4)$$
$$\text{Rebolledo.} \quad D = 1.15 + 0.035\,(S - V) \qquad (5)$$

In these, D = depth of key; S = span; R = radius of curve of intrados; V = versine or rise. If the intrados be elliptical, R = assumed or approximate radius at crown. All dimensions in feet.

Of these formulæ (1) is said by Leville to apply to all curves of intrados, semi-circular, segmental, or elliptical. Formula (2) applies to semi-circular arches, and also to segmental ones, or those formed by an arc of a circle less than a semi-circle, when the rise is more than $\dfrac{S}{6}$. If less, (3) applies. This formula is slightly changed from the original.

It will be instructive to apply these formulæ to a series of arches of different spans, for the purpose of comparison. Commencing with semi-circular arches of 30, 50 and 100 ft. span, the following table gives the value of D from formulæ (1); (2); (4) and (5).

SEMI-CIRCULAR ARCHES.

S	D			
	(1)	(2)	(4)	(5)
30	2.05	1.97	2.33	1.68
50	2.75	2.40	2.76	2.03
100	4.50	3.18	3.48	2.90

For segmental arches of 60°, of which the radius is equal to the span, and the versine or rise is 0.134 of span or radius, formulæ (1), (3), (4) and (5) give:

SEGMENTAL ARCHES OF 60°.

S	D			
	(1)	(3)	(4)	(5)
30	2.05	2.31	2.33	2.06
50	2.75	2.83	2.76	2.67
100	4.50	3.80	3.48	4.18

These results agree better than the previous ones, but it is evident that we can exercise considerable latitude in the matter of keys and yet have good authority, and still better, good examples to sustain us.

These thicknesses of key apply more particularly to railroad and highway bridges. They are sufficient to carry on the arch ring alone and without taking account of the spandril backing, track and trains, with 2 or 3 ft. of earth filling over the extrados. For heavier embankments some authors recommend an addition to the depth of key of 2 per cent. of height of embankment. Thus, if an arch carries a 50-ft. embankment, add one foot to depth of key.

In the absence of special empirical formulæ for aqueduct arches, the above may be safely taken for this purpose, especially since for aqueducts the spandril backing will generally be carried up level with the extrados at the crown.

It is a well-known fact that a semi-circular arch can be carried up to a considerable height above the spring line before the voussoirs begin to bear upon the centering, as they are kept in place below this height, by friction. It has been found that the point at which centering becomes necessary to support the voussoirs of a semi-circular arch occurs at about half the height of the rise, or at 60° measured from the vertical. The joint at this point, or the point nearest to it, is known as the "joint of rupture," and is one of the most important elements of arch designing. It is up to this level that the backing of the haunches should always be carried. The arch proper commences at this joint, and includes 120°. All below this joint must be considered as forming part of the abutments.

In a well-proportioned voussoir arch, the thickness of the arch ring should increase from the crown toward the haunches. The radial length of any joint between the key and the joint of rupture should be such that its vertical projection, or the cosine of the angle which it makes with the vertical, shall be equal to the depth, D, of the key. It is, therefore, $= \dfrac{D}{\cos.\ \alpha}$, α being the angle which the joint makes with the vertical. The cosine of 60° (or sine of 30°) being ½, we have for the length, L, of the joint of rupture of a semi-circular arch :

$$L = 2D \qquad (6)$$

The length of any joint intermediate between the key and joint of rupture may be found as shown on the right-hand side of Fig. 24, by drawing the horizontal line df at the distance $cd =$

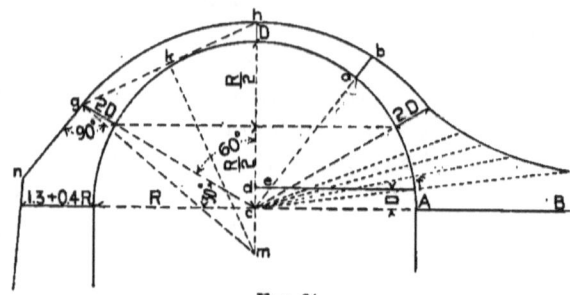

FIG. 24.

D from the springing line $A B$. Then the length of any joint, $a b$, is found by drawing $c b$, and taking $e b = R$.

This process can be continued below the joint of rupture, as shown in the figure, when the curve of the extrados rapidly flattens, becoming finally an asymptote to the springing line.

The above process is due to Dejardin. Dubosque gives a more rapid method, resulting in a somewhat greater thickness, which is shown on the left-hand side of Fig 24, and is described as follows: Join g and h, the exterior extremities of the joint of rupture and the imaginary joint at the crown. Bisect $g\,h$ in K. Draw $K\,m$ perpendicular to $g\,h$, intersecting the vertical at m. From m, with radius $m\,g$, $m\,h$, describe the arc $g\,h$ which is the required extrados. Dubosque finishes the extrados by drawing the tangent $g\,n$ to intersect with the back of the abutment, produced, as shown. This process applies to all arches, whether semi-circular, segmental, or elliptical.

Since in a semi-circular arch the joint of rupture is situated at an angle of 60° from the vertical, it follows that all segmental arches of which the amplitude is equal to or less than 120°, or in other words in which quotient of the span divided by the rise is equal to or greater than 3.46, have their joint of rupture at the springing line. Since the length of any joint is $\dfrac{D}{\cos\alpha}$, the length of skewback joint of such a segmental arch can be found by multiplying the depth of key by the radius of intrados, and dividing it by the radius minus the rise, thus:

$$L' = \frac{DR}{R-V}$$

Since the radius of a circular arc is given by the relation:

$$R = \frac{S^2 + 4V^2}{8V}$$

we have:

$$L' = \frac{S^2 + 4V}{S^2 - 4V^2} \qquad (7)$$

If the segmental arch has an amplitude of more than 120°, that is if $\frac{S}{V} < 3.46$ the position of the joint of rupture is, of course, at the distance of half the radius from the crown, the same as for a semi-circular arch.

For elliptical or false elliptical arches the joint of rupture occurs at half the rise, the same as for semi-circular ones, but its length is differently determined. Croizette-Desnoyers has given the following series of coefficients for such arches for different values of the ratio $\frac{S}{V}$:

$$\frac{S}{V} = 3; \; L = 1.80 \, D \qquad (8)$$

$$\frac{S}{V} = 4; \; L = 1.60 \, D \qquad (9)$$

$$\frac{S}{V} = 5; \; L = 1.40 \, D \qquad (10)$$

For other ratios L can be found by interpolation :

Arches of this class are generally of the "false-elliptical" type, that is, are formed by an odd number of circular arcs, tangent to each other, the resultant curve closely approximating a true ellipse, three being the smallest number of such arcs that can be used. This is called a "three-centred" arch, and is suitable to ratios not greater than $\frac{S}{V} = 3$. When the value of the ratio exceeds 3 a greater number of centres should be used.

The two radii, in terms of span and rise, of a three-centred arch are given by Dubosque :

$$R = \frac{S}{2} + 0.683 \left(S - 2\,V \right) \qquad (11)$$

$$r = S - R = \frac{S}{2} - 0.683\left(S - 2V\right) \qquad (12)$$

In which R = radius of large central arc, and r = that of the two smaller end arcs.

As regards top thickness or thickness at the spring line, T, of abutments, a good general formula is given by Boix :

$$T = 1.30 + 0.11\, S \sqrt{\frac{S}{V}} \qquad (13)$$

For semi-circular arches, this reduced to :

$$T' = 1.30 + 0.20\, S. \qquad (14)$$

Formulæ (13) and (14) take no account of the height of the abutment, which is not, contrary to what might be supposed, a very important factor. If the top thickness be determined by the formula, the batter of the wall will amply provide for any additional thickness rendered necessary by the greater or less height.

A good formula, of, I believe, German or Russian origin, for thickness of abutments of semi-circular arches in which the height, H, enters is :

$$T = 1 + 0.20\, S + 0.16\, H. \qquad (15)$$

In ordinary cases this may be reduced to :

$$T = 0.30\, S.$$

For piers, Rebolledo gives :

$$T = 2.50\, D + 0.10\, H.$$

In which D = depth of key of arch, and H = height of pier. In wide spans and heavy loads, the weight borne by foot of piers and abutments must be considered with regard to resistance to crushing.

EXAMPLE.—Determine the principal dimensions of a 3-centred arch, 30-ft. span, 10-ft. rise, abutments 12 ft. high. See Fig. 25.

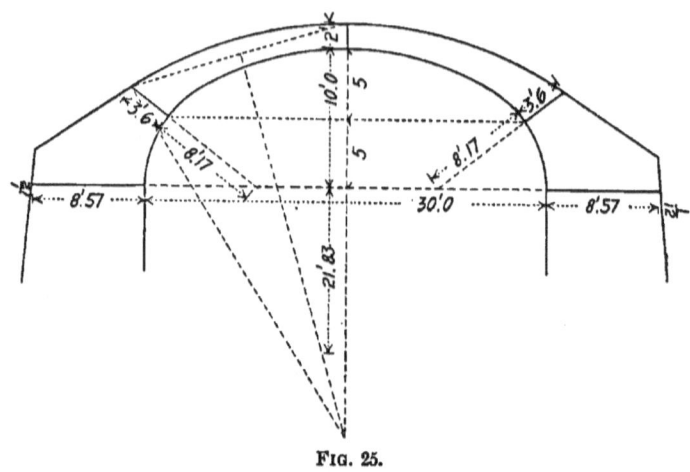

Fig. 25.

From (11) or (12).

$R = 21.83$ ft.
$r = 8.17$ ft.

Thickness, D, of key from formulæ (1) to (5) respectively, 2.05; 2.27; 2.33; 1.85. Take $D = 2$ ft. Joint of rupture occurs at $\dfrac{V}{2} = 5$ ft. below crown, and its length, per (8) $= 1.80\,D = 3.60$ ft. The top thickness, or thickness at spring-line of abutments per (13) $= 1.30 + 0.14 \times 30\sqrt{\dfrac{30}{10}} = 8.57$ ft.

The formulæ for thickness of abutments give that necessary to sustain the thrust of the arch. No account is taken of the counter thrust of the earth embankment behind the abutment. In many cases the abutments could be lightened by relieving arches or otherwise.

When an arch fails, it is generally on account of settling or spreading of the abutments, or to bad work or materials. If there is no movement in the supports, it can only fail by direct crushing of the materials—a very rare case—or by distortion of the arch. Distortion is caused by a sinking at one point and rising at another. The object is, therefore, to get such a substantial and evenly distributed permanent load upon the arch, if possible, before striking the centres, that no irregular strains can come upon it to cause distortion. Naturally, the lighter the permanent road, the stronger and stiffer must be the arch in order to resist transient and unequal loading. These remarks apply more particularly to viaducts, for an aqueduct arch is seldom subjected to unbalanced stresses.

As regards the best form of arch, a given opening can be successfully spanned by an arch of any form. Generally speaking, the curve of intrados is selected in reference to the special conditions of each case, the amount of head room required, etc. There are, however, certain forms which, other things being equal, are best fitted for certain kinds of loading. If an arch is to sustain a single load, concentrated at the centre and heavy as compared with the weight of the arch itself, the gothic arch is most suitable. If the loading increases gradually from the crown to the haunches, as in the case of an ordinary earthen embankment, the curve of pressure would more nearly approach the arc of a circle. Should the loading be much greater at the haunches than at the crown, the curve would approach the elliptical form. In all cases the curve will be found to *rise to the pressure*; for the true curve of pressure is always represented, inverted, by a flexible cord, similarly loaded to the arch. Where head room near haunches can be spared, and great economy of material is not essential, the full centred, or semi-circular arch, will generally be preferred, both for its great structural stability and the beauty of its proportions. The "High Bridge" of the old Croton aqueduct may be here instanced. When the full centred arch is inadmissible, the segmental arch, or that formed by the arc of a circle less than 180° in amplitude, has much to recommend it. Of these, the arc of 60°, with span equal to radius and rise = 0.134 S seems a very happy selection both as to appearance and convenience of dimensions. Should an arch be intended to support a body of water in direct contact with its extrados, the proper theoretical form would be that of the hydrostatic arch, which resembles a cycloid, which in turn resembles an ellipse. In all practical cases, however, such as tanks or aqueducts, a level bottom is necessary, and the spandrils are built up level with the crown. In this case the conditions governing the hydrostatic curve do not obtain, and the load from the water is simply an equally distributed one, pressing at all points vertically downward, or at least practically so.

Although the above empirical formulæ suffice to correctly proportion any except very unusual forms, it will be well to devote a few words to the more theoretical features of the subject.

Suppose it were wished to calculate the curve of pressure in the semi-circular voussoir arch, half of which is shown in Fig. 26.

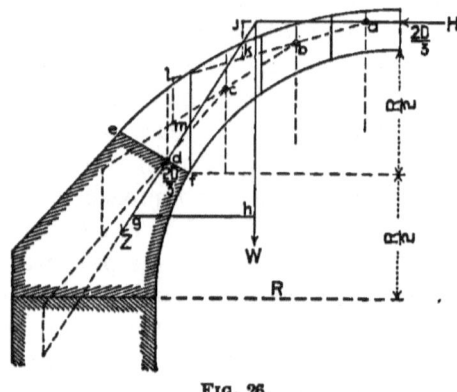

FIG. 26.

Assuming it to be of good materials and workmanship, without which all calculation would be impossible, that part of the structure lying above the joint of rupture ef is taken as comprising the arch proper. The section is supposed to be divided into an arbitrary number of fictitious voussoirs a, b, c and d, in the figure, which will give, equally well with the true ones, the form of the curve of pressure. The weight of and upon each of these voussoirs is then estimated, and the position of the line passing through the centre of gravity of each voussoir and its load determined. From these data the position of the vertical line W passing through the centre of gravity of the entire section above joint of rupture is fixed. The horizontal line H cutting the central joint at one third $\left(\dfrac{=D}{3}\right)$ of its length from the top is then drawn to its intersection with the vertical W. From the point of intersection the line Z is drawn cutting the joint of rupture at one third $\left(\dfrac{=2D}{3}\right)$ of its length from the bottom. The weight of the half arch and load above ef is then laid off to scale on W, and from its extremity the horizontal line gh is drawn to Z, completing the triangle of forces. The length of the line gh to the same scale as W gives the value of horizontal thrust at the crown. On the horizontal line H produced, take $aj = gh$, and from it extremity lay off jK = weight of voussoir, a, and load. Join ak. On ak produced, lay off $bl = ak$. From l, lay off lm = weight of voussoir, b, and load. Proceed thus, working the curve down from voussoir to voussoir to the line of rupture. The last resultant should coincide with the oblique line Z, which fact furnishes an excellent check upon the accuracy of the work. The broken line representing the line of pressure can now be harmonized with a curve, drawn by hand and the actual voussoirs laid down on the drawing to show where and at what angle the curve cuts their joints.

The line of pressure may be continued down to the springing line and on through the abutment, but this is not generally necessary.

The reason why the points of application of H and Z at the crown and joint of rupture are placed at the upper and lower extremities, respectively, of the middle third of these joints is the following: The general tendency of all arches, except those carrying very unusual loads at the haunches, is to sink at the crown and consequently rise at the haunches. When the crown sinks the arch opens at the intrados, in the neighborhood of the key, rotating upon its upper edge, at the extrados. Inversely, the joint of rupture will open at the extrados, rotating around its lower edge, at the intrados. If the joint at the crown does not actually open, there is always a *tendency* to do so, and a corresponding tendency at the joint of rupture; and at the points around which the voussoirs *tend* to rotate the compressive stresses are at their maximum, diminishing progressively until reaching the other extremity of the joint, where the tendency is to open. At this point the compressive stresses become zero. The total compressive stresses on these joints may therefore be represented graphically by the area of a right-angle triangle constructed upon each joint, with its base at the extremity around which rotation tends to take place, and its apex at the extremity which tends to open. The resultant stress passes through the centre of gravity of each triangle, which is at one-third of its height from the base. The heights of the triangles being represented by the length of the joints, the points of application of H and Z must, to conform with the above theory, be placed as in the figure. In this connection see pages 87 and 88.

For a properly proportioned arch, such as would be the outcome of the practical rules already laid down, and is shown in Fig. 26, the amount of horizontal thrust at the crown—which is to arch calculation what abutment reaction is to that of girders and bridges—can be obtained more readily than in the above example by the use of Navier's formula:

$$H = P \times R \qquad (17)$$

in which P = pressure or weight per square unit at the crown, and R = radius of the intrados at the crown, expressed in the same linear unit. Thus, in last example, Fig, 26, if the weight of arch and load on or in the immediate vicinity of the key were 1,000 lbs. per square foot, and the radius 15 ft., the total horizontal thrust would = 15,000 lbs. In example, Fig. 25, if the unit pressure were the same, the radius at the crown being 21.84 ft , the total horizontal thrust H would equal 21,840 lbs.

As regards the whole subject of arch design, it may be broadly stated, on the authority of Dejardin, that if the arch is proportioned according to the rules already laid down, and is well constructed with good materials, no calculation whatever is needed to demonstrate its stability. If, however, the engineer should be called upon to discuss an existing structure built upon different lines (and which, nevertheless, might fulfil all the requirements of stability) he should proceed as above directed, making different assumptions until a line of pressure is found which shall at least lie inside of the arch, at all points, the presumption being that if such a line can exist, the arch will find it. Should no curve be found which did not cut the intrados or extrados at some point or points, it would indicate that rotation would occur around such point or points, with intense local compressive stress. If this result were found in any design under discussion, it would justify the rejection of such design. If it were found in any existing structure, it would prove that the construction and materials of the arch were of a nature to permit it to resist bending moments at such points, or that the assumed data regarding weights and loading were incorrect. Should the line of pressure, while not leaving the limits of the section, approach very near to the intrados or extrados, the amount of compressive stress set up at these points is to be determined according to the rules laid down for masonry dams, considering the compression to be concentrated upon the area of joint lying between the point of application of the pressure and the nearer extremity of the joint.

The chief uncertainty which militates against the value of all arch calculation lies in the fact that we are obliged in most cases to make almost random guesses at the superincumbent loads actually and not theoretically sustained by the arch. As a simple example of this may be cited the case of an arch sustaining a high masonry wall. Theoretically, the whole weight of the wall would be resting upon the arch, and the higher the wall the more danger to the arch. Practically, we know that the higher the wall the less danger would there be of its falling in if we should break an opening through it near the bottom.

It will be noticed in all that precedes that the joint of rupture has been placed at half the height of the rise. It is found in practice that this assumed position results in a well-proportioned and secure arch, particularly if the masonry backing be carried up to this height at the extrados. Strictly speaking, however, we cannot so broadly generalize the problem, and probably it is impossible to tell exactly where the joint of rupture is located in any existing arch. Lamé and Clapeyron assert that

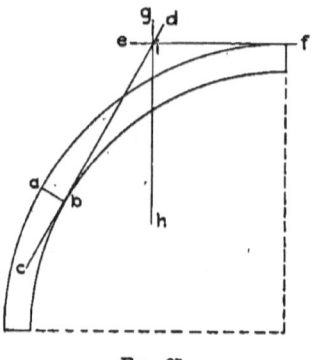

FIG. 27.

the line of rupture ab, Fig. 27, is so situated that the tangent cd, at the point b, and the tangent ef at the crown, will always intersect at the line $g\,h$, passing through the centre of gravity of the mass lying above the line of rupture in such a position that these three lines will all meet at one and the same point, *i.* To determine the position of the joint of rupture by this rule, it would be necessary to operate by trial and error. It might be used to test the assumed position of the joint in any particular case, but it does not seem to possess sufficient practical utility to render its use recommendable.

All that precedes relates to the voussoir arch, the fundamental principle of which is that it preserves its stability by equilibrium alone, independent of any cohesion of the mortar in which the stones are bedded. Brick and concrete arches belong to a totally different class, as their stability depends principally or wholly upon the cohesion of the mortar. They are monolithic in character, and while the empirical formulæ established for voussoir arches can be used to determine their proper dimensions, the calculations respecting the line of pressure do not apply.

The increased thickness toward the haunches of a brick arch may be given by bonding in additional rings, or, as will generally be found preferable, by giving the arch a uniform thickness throughout, and obtaining the desired increase by adding to the spandril backing, as shown in Fig. 28. In the case of brick arches of small span, it is best to build them in independent concentric rings, but if the span and radius of curvature are relatively large, it will frequently be found advisable to bond the rings together.

Should it be desired to apply calculation to a brick arch in order to ascertain its probable stability, recourse should be had to the old method of Lamé and Clapeyron, which is based upon Boistard's experimental researches. Without going into a full description of this method, which can be found in text books, the formula will be given to determine whether a given arch is stable as against sinking at the crown and raising at the haunches, by far the most common case of failure.

Referring to Fig. 28, supposing the abutments to be unyielding, and considering only the arch $a\,b\,c\,d\,e\,f$, exclusive of the

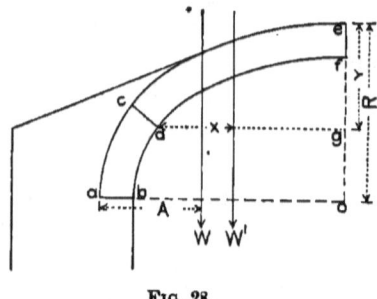

FIG. 28.

backing, the notation is as follows, $a\,b$ being the springing line and $c\,d$ the assumed joint of rupture:

W = weight of mass $a\,b\,c\,d\,e\,f$ including loading.
W' = weight of mass $c\,d\,e\,f$ including loading.
A = distance from a to line passing through centre of gravity of W.
x = distance from d to line passing through centre of gravity W'.
y = vertical distance $e\,g$, from d to e.
R = total rise, $o\,e$, from springing line to extrados.

Then the equation for exact static equilibrium is:

$$WA - W'R\frac{x}{y} = 0 \qquad (18)$$

and for stability:

$$WA - W'R\frac{x}{y} > 0 \qquad (19)$$

Thus, in the arch shown in Fig. 28, assuming a span of 30 ft.; a rise of 10 ft., and a depth of key = 2.50 ft., let the data be:

$W = 6{,}000$ lbs.
$W' = 2{,}250$ lbs.
$R = 12.50$ ft.
$A = 7.40$ ft.
$x = 5.40$ ft.
$y = 7.50$ ft.

Then from (18) and (19):

$$6000 \times 7.40 - 2250 \times 12.50 \times \frac{5.4}{7.5} = 24,150 \text{ lbs.}$$

This would indicate that the moment WA, making for stability, had a factor of safety of nearly 1.85.

It will be noted that equations (18) and (19) contain two variables, depending upon the position of the joint of rupture. Write (18) in this form:

$$R \left(\frac{WA - W' x}{R} \frac{x}{y} \right) = 0 \qquad (20)$$

Then, should the arch break, it would be at some joint, $a\, b$, such as would make $\frac{W' x}{y}$ maximum. This point can be found by trial and error. If such maximum value gives a negative result in (20), it would indicate that the portion of the arch below the joint of rupture was too light.

It is necessary that the constructing engineer should be familiar with the above processes of calculation, or, at least, be aware of their existence, in order to know what can and cannot be accomplished by figuring. The successful development of arch building having been mainly along purely experimental lines, very little aid is to be hoped for from abstract mathematical reasoning, for the very good reasons—among others—that the most important data must be assumed, or, in other words, guessed at, and that in the case of voussoir arches the adhesion and cohesion of the mortar is and must be ignored, although it may really play a very important part, particularly where small materials are used. In the case of concrete arches, when the limit of small and amorphous materials is reached, the strength of the mortar is everything.

The first requisites for an arch are unyielding foundations and abutments, without which fracture and perhaps destruction must ensue. Then come good workmanship and materials, and great judgement as to time and manner of striking centres and loading arch. *Design* has been purposely left to the last, because if all the other requirements are observed, the shape and dimensions of the mere arch ring are matters which, though important, are still of lesser moment.

For further details of arch design, see "Van Nostrand's Engineering Magazine," for December, 1883, and February, 1884.

EXPLANATION OF THE TABLES.

Table 1 gives the areas of circles in square feet, corresponding to diameters in inches.

Table II contains five columns, headed respectively D, V, Q, G and $H\,P$, giving various properties of rough cast-iron long pipes of different diameters having falls of one to ten per thousand. D = diameter of pipe in inches; V = velocity in feet per second; Q = discharge in cubic feet per second; G = discharge in U. S. gallons per hour, and $H\,P$ = theoretical or net horse-power necessary to raise the quantity discharged one foot high. In this table V is calculated by the formula (1 *bis.*) or (1 *ter.*), Q is calculated by multiplying V by the area of the pipe, G by multiplying Q by 27,000, and $H\,P$ either by dividing Q by 8.82 or multiplying G by 42 and pointing off seven decimal places.

EXAMPLE.—A pipe 40 inches in diameter is laid to a grade of $\frac{3}{1000}$. What is the discharge and the net horse-power necessary to lift the volume discharged to a height of 113 ft.? The discharge is 947,241 U. S. gallons per hour, and the horse-power for each foot of lift is 3.98. Then $3.98 \times 113 = 449.74$ is the required net horse-power.

Any of the above data can be obtained for falls per thousand not given in the table by multipiying the values given for falls of $\frac{1}{1000}$ by the square root of the fall.

EXAMPLE.—A pipe 20 in. in diameter has a fall of 3.43 per

1,000. What is the discharge in cubic feet per second? What the horse power per foot of lift? For a fall of $\frac{1}{1000}$, $Q = 3.511$. Therefore with a fall of 3.43, $Q = 3.511\sqrt{3.43} = 6.50$ cu. ft. per second. Again, for $\frac{1}{1000}$, $HP = 0.40$. Therefore, for $\frac{3.43}{1000}$, $HP = 0.40\sqrt{3.43} = 0.74$.

Table III. gives the value of $\dfrac{RI}{U^2}$ for different values of R, from which the mean velocity U can be deduced.

EXAMPLE.—Fig. 29 represents the cross-section of the old

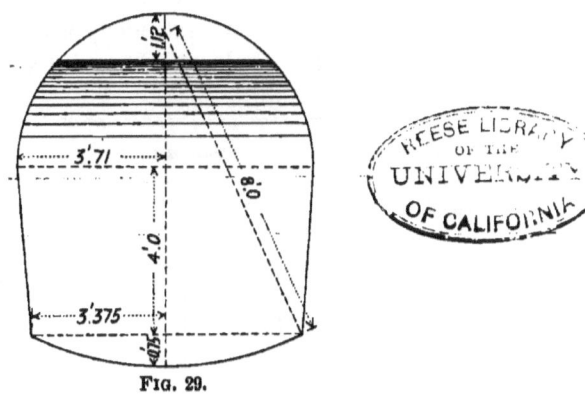

FIG. 29.

Croton Aqueduct. When running to within 1.12 ft. of the crown, as shown in the figure, the wet section = 49.19 sq. ft., the wet perimeter 20.72 ft., and the mean hydraulic radius

$R = \dfrac{49.19}{20.72} = 2.37$. Let $I = 0.00021$. Then by table III.:

$$\dfrac{RI}{U^2} = 0.00006794.$$

Also by the data:

$$\frac{R\,I}{U^2} = \frac{0.0004977}{V^2}.$$

Therefore:

$$\frac{0.0004977}{U^2} = 0.00006794$$

$$U^2 = \frac{49770}{6794}$$

$$U = 2.71.$$

This value agrees quite nearly with that obtained by experiments with floats executed under the author's direction in 1884.

It will be readily seen how extremely useful this table is in rapidly determining the discharge of conduits.

TABLE I.

D = inches.	A = square feet.	D = inches.	A = square feet.
1	0.00545	26	3.6868
2	0.02180	27	3.9760
3	0.0491	28	4.2760
4	0.0872	29	4.5868
5	0.1364	30	4.9087
6	0.1964	31	5.2413
7	0.2672	32	5.5848
8	0.3490	33	5.9394
9	0.4418	34	6.3048
10	0.5454	35	6.6812
11	0.6599	36	7.0086
12	0.7854	37	7.4665
13	0.9217	38	7.8756
14	1.0690	39	8.2957
15	1.2272	40	8.7264
16	1.3962	41	9.1682
17	1.5762	42	9.6211
18	1.7671	43	10.0847
19	1.9689	44	10.5589
20	2.1816	45	11.0440
21	2.4052	46	11.5408
22	2.6397	47	12.0479
23	2.8852	48	12.5660
24	3.1416	49	13.0951
25	3.4087	50	13.6354

WATER SUPPLY ENGINEERING. 157

TABLE II.

$$\frac{1}{1000}$$

D	V	Q	G	HP
3	0.56	0.027	729	0.003
4	0.66	0.057	1539	0.007
6	0.83	0.163	4401	0.02
8	0.99	0.344	9288	0.04
10	1.12	0.612	16524	0.07
12	1.23	0.966	26082	0.11
14	1.34	1.432	38064	0.16
16	1.44	2.010	54270	0.23
18	1.53	2.704	73008	0.31
20	1.61	3.511	94797	0.40
22	1.70	4.488	121176	0.51
24	1.77	5.561	150147	0.63
26	1.84	6.784	183168	0.77
28	1.91	8.167	220509	0.93
30	1.99	9.769	263763	1.11
32	2.06	11.505	310635	1.31
34	2.12	13.367	360909	1.52
36	2.20	15.552	419904	1.76
38	2.26	17.800	480600	2.02
40	2.32	20.247	546669	2.30
42	2.38	22.898	618246	2.60
44	2.43	25.658	692766	2.91
46	2.49	28.737	775899	3.26
48	2.54	31.918	861786	3.62

$$\frac{2}{1000}$$

D	V	Q	G	HP
3	0.79	0.039	1053	0.005
4	0.93	0.081	2187	0.010
6	1.18	0.231	6237	0.03
8	1.40	0.489	13203	0.06
10	1.59	0.868	23436	0.10
12	1.74	1.366	36882	0.16
14	1.90	2.031	54837	0.23
16	2.05	2.862	77274	0.33
18	2.16	3.817	103059	0.43
20	2.28	4.973	134271	0.56
22	2.40	6.336	171072	0.72
24	2.50	7.855	212085	0.89
26	2.60	9.586	258822	1.09
28	2.70	11.545	311715	1.31
30	2.81	13.794	372438	1.56
32	2.91	16.252	438804	1.84
34	3.00	18.915	510705	2.15
36	3.11	21.985	593595	2.49
38	3.20	25.203	680481	2.86
40	3.28	28.625	772875	3.25
42	3.37	32.423	875421	3.68
44	3.44	36.323	980721	4.12
46	3.52	40.624	1096848	4.61
48	3.59	45.112	1218024	5.12

TABLE II.—(CONTINUED.)

$$\frac{3}{1000}$$

D	V	Q	G	H P
3	0.97	0.048	1296	0.006
4	1.15	0.100	2700	0.012
6	1.44	0.282	7614	0.03
8	1.72	0.600	16200	0.07
10	1.95	1.065	28755	0.12
12	2.13	1.672	45144	0.19
14	2.32	2.480	66960	0.28
16	2.51	3.504	94608	0.40
18	2.65	4.683	126441	0.53
20	2.79	6.085	164295	0.69
22	2.93	7.735	208845	0.88
24	3.05	9.615	259605	1.09
26	3.19	11.761	317547	1.33
28	3.31	14.154	382158	1.61
30	3.45	16.936	457272	1.92
32	3.57	19.938	538326	2.26
34	3.67	23.139	624753	2.62
36	3.81	26.933	727191	3.05
38	3.91	30.795	831465	3.49
40	4.02	35.083	947241	3.98
42	4.12	39.639	1070253	4.50
44	4.21	44.453	1200231	5.04
46	4.31	49.742	1343034	5.64
48	4.40	55.290	1492830	6.27

$$\frac{4}{1000}$$

D	V	Q	G	H P
3	1.12	0.055	1485	0.006
4	1.32	0.115	3105	0.013
6	1.67	0.327	8829	0.04
8	1.98	0.691	18657	0.08
10	2.25	1.230	33183	0.14
12	2.46	1.931	52137	0.22
14	2.68	2.865	77355	0.33
16	2.90	4.048	109296	0.46
18	3.06	5.407	145989	0.61
20	3.22	7.023	189621	0.80
22	3.40	8.976	242352	1.02
24	3.54	11.123	300321	1.26
26	3.68	13.568	366336	1.54
28	3.82	16.334	441018	1.85
30	3.98	19.538	527526	2.22
32	4.12	23.010	621270	2.61
34	4.24	26.733	721791	3.03
36	4.40	31.104	839808	3.53
38	4.52	35.600	961200	4.04
40	4.64	40.493	1093311	4.59
42	4.76	45.796	1236492	5.19
44	4.86	51.317	1385559	5.82
46	4.98	57.474	1551798	6.52
48	5.08	63.835	1723545	7.24

WATER SUPPLY ENGINEERING.

TABLE II.—(CONTINUED.)

$$\frac{5}{1000}$$

D	V	Q	G	H P
3	1.25	0 061	1647	0.007
4	1.48	0 129	3483	0.015
6	1.86	0.365	9855	0.04
8	2.23	0.778	21006	0.09
10	2.51	1.370	36990	0.16
12	2.75	2.159	5293	0.25
14	3.00	3 207	86589	0.36
16	3.24	4.523	122121	0.51
18	3.42	6.043	163161	0.69
20	3.60	7.852	212004	0.89
22	3.79	10.006	270162	1.14
24	3.96	12.442	335934	1.41
26	4.11	15.154	409158	1.72
28	4.27	18.259	492993	2.07
30	4.45	21.845	589815	2.48
32	4.61	25 747	695169	2.92
34	4 74	29.886	806922	3.39
36	4 92	34 779	939033	3.94
38	5 05	39.774	1073898	4.51
40	5.19	45.293	1222911	5.14
42	5.32	51.184	1381968	5.80
44	5.43	57.335	1548045	6.50
46	5.57	64 283	1735641	7.20
48	5 68	71 375	1927125	8.09

$$\frac{6}{1000}$$

D	V	Q	G	H P
3	1.37	0.067	1800	0.008
4	1.62	0 141	3807	0.016
6	2.04	0.400	10800	0.05
8	2.44	0.852	23004	0.10
10	2.75	1.502	40554	0 17
12	3.01	2.363	63801	0.27
14	3.28	3.506	94662	0.40
16	3.55	4.956	133812	0.56
18	3.75	6.626	178902	0.75
20	3.94	8.593	32011	0.98
22	4.15	10.956	295812	1.24
24	4.33	13.605	367335	1.54
26	4.50	16.592	447984	1.88
28	4.68	20.012	540324	2.27
30	4.87	23.907	645489	2.71
32	5.04	28.148	759996	3.19
34	5.19	32.723	883521	3.71
36	5.39	38.102	1028754	4.32
38	5.53	43.554	1175958	4.94
40	5.68	49.569	1338363	5.62
42	5.83	56.09	1514430	6.36
44	5.95	62.826	1696302	7.13
46	6.10	70.400	1900800	7.98
48	6.22	78.161	2110347	8.86

WATER SUPPLY ENGINEERING.

TABLE II.—(CONTINUED.)

$\dfrac{7}{1000}$

D	V	Q	G	H P
3	1.48	0.073	1971	0.008
4	1.75	0.152	4104	0.017
6	2.20	0.431	11637	0.05
8	2.62	0.914	24678	0.10
10	2.97	1.619	43713	0.18
12	3.26	2.559	69093	0.29
14	3.55	3.795	102465	0.43
16	3.84	5.361	144747	0.61
18	4.05	7.156	193212	0.81
20	4.26	9.295	250065	1.05
22	4.48	11.827	319329	1.34
24	4.68	14.705	397035	1.67
26	4.87	17.956	484812	2.04
28	5.05	21.594	583038	2.45
30	5.27	25.870	698490	2.93
32	5.45	30.438	821826	3.45
34	5.61	35.371	955017	4.01
36	5.82	41.142	1110834	4.67
38	5.98	47.098	1271646	5.34
40	6.14	53.578	1446606	6.08
42	6.30	60.612	1636524	6.87
44	6.43	67.894	1833138	7.70
46	6.58	75.933	2050191	8.61
48	6.72	84.444	2279988	9.58

$\dfrac{8}{1000}$

D	V	Q	G	H P
3	1.58	0.077	2079	0.009
4	1.87	0.163	4401	0.019
6	2.36	0.463	12501	0.05
8	2.80	0.977	26379	0.11
10	3.18	1.733	46791	0.20
12	3.48	2.732	73764	0.31
14	3.79	4.052	109404	0.46
16	4.10	5.724	154548	0.65
18	4.33	7.651	206577	0.87
20	4.55	9.928	268056	1.13
22	4.79	12.646	341442	1.43
24	5.00	15.710	424170	1.78
26	5.20	19.172	517644	2.17
28	5.40	23.000	623130	2.62
30	5.63	27.638	746226	3.13
32	5.83	32.561	879147	3.69
34	6.00	37.830	1021410	4.29
36	6.22	43.969	1187163	4.99
38	6.39	50.328	1358856	5.71
40	6.56	57.243	1545561	6.49
42	6.73	64.749	1748223	7.34
44	6.87	72.540	1958580	8.23
46	7.04	81.249	2193723	9.21
48	7.18	90.224	2436048	10.23

WATER SUPPLY ENGINEERING. 161

TABLE II.—(CONTINUED.)

$$\frac{9}{1000}$$

D	V	Q	G	H P
3	1.68	0.082	2214	0.01
4	1.98	0.172	4644	0.02
6	2.50	0.490	13230	0.06
8	2.97	1.037	27999	0.12
10	3.37	1.837	49599	0.21
12	3 69	2.897	78219	0 33
14	4.02	4.297	116019	0.49
16	4.33	6.045	163215	0.69
18	4.59	8.111	218997	0.92
20	4.83	10 539	284553	1.20
22	5.09	13.438	362836	1.52
24	5.30	16.653	449631	1.89
26	5 52	20.352	549504	2.31
28	5.73	24.501	661527	2.78
30	5.97	29 307	791289	3.32
32	6.18	34.515	931905	3 91
34	6.36	40.100	1082700	4.55
36	6.60	46.655	1259685	5.29
38	6.78	53.400	1441800	6.06
40	6.96	60.733	1639791	6.89
42	7.14	68.694	1854738	7.79
44	7 29	76.975	2078325	8.73
46	7.47	86.211	2327697	9 78
48	7.62	95.753	2585331	10 86

$$\frac{10}{1000}$$

D	V	Q	G	H P
3	1.77	0.087	2349	0.01
4	2 09	0.182	4914	0.02
6	2.63	0.515	13905	0.06
8	3.16	1.103	29781	0.13
10	3.55	1.935	52245	0 22
12	3.89	3.054	82458	0.35
14	4.24	4.533	122391	0.51
16	4.58	6.394	172638	0.73
18	4.84	8.552	230904	0.97
20	5.09	11.106	299862	1.26
22	5.35	14.124	381348	1.60
24	5.60	17.595	475065	2.00
26	5.82	21.458	579366	2.43
28	6.04	25.823	697221	2.93
30	6.29	30.878	833706	3.50
32	6.51	36.358	981666	4.12
34	6.70	42.244	1140588	4.79
36	6.96	49.200	1328400	5.58
38	7.15	56.313	1520451	6.39
40	7.34	64.049	1729323	7.26
42	7.53	72.446	1956042	8.22
44	7.68	81.093	2189511	9.20
46	7.87	90.828	2452356	10.30
48	8.03	100.905	2724435	11.44

WATER SUPPLY ENGINEERING.

TABLE III.

R	$\dfrac{R\,l}{U^2}$	R	$\dfrac{R\,l}{U^2}$	R	$\dfrac{R\,l}{U^2}$
0.05	0.00015800	1.75	0.00006863	4.0	0.00006715
0.10	0.00011200	1.80	0.00006856	4.1	0.00006712
0.15	0.00009607	1.85	0.00006849	4.2	0.00006710
0.20	0.00008900	1.90	0.00006842	4.3	0.00006707
0.25	0.00008440	1.95	0.00006836	4.4	0.00006705
0.30	0.00008133	2.00	0.00006830	4.5	0.00006702
0.35	0.00007914	2.05	0.00006824	4.6	0.00006700
0.40	0.00007750	2.10	0.00006819	4.7	0.00006698
0.45	0.00007622	2.15	0.00006814	4.8	0.00006696
0.50	0.00007520	2.20	0.00006809	4.9	0.00006694
0.55	0.00007437	2.25	0.00006804	5.0	0.00006692
0.60	0.00007367	2.30	0.00006800	5.1	0.00006690
0.65	0.00007308	2.35	0.00006796	5.2	0.00006689
0.70	0.00007257	2.40	0.00006792	5.3	0.00006687
0.75	0.00007213	2.45	0.00006788	5.4	0.00006685
0.80	0.00007175	2.50	0.00006784	5.5	0.00006684
0.85	0.00007141	2.55	0.00006780	5.6	0.00006682
0.90	0.00007111	2.60	0.00006777	5.7	0.00006681
0.95	0.00007084	2.65	0.00006774	5.8	0.00006679
1.00	0.00007060	2.70	0.00006770	5.9	0.00006678
1.05	0.00007038	2.75	0.00006767	6.0	0.00006677
1.10	0.00007018	2.80	0.00006764	6.1	0.00006675
1.15	0.00007000	2.85	0.00006761	6.2	0.00006674
1.20	0.00006983	2.90	0.00006759	6.3	0.00006673
1.25	0.00006968	3.0	0.00006753	6.4	0.00006672
1.30	0.00006954	3.1	0.00006748	6.5	0.00006671
1.35	0.00006941	3.2	0.00006744	7.0	0.00006666
1.40	0.00006929	3.3	0.00006739	7.5	0.00006661
1.45	0.00006917	3.4	0.00006735	8.0	0.00006658
1.50	0.00006907	3.5	0.00006731	8.5	0.00006654
1.55	0.00006897	3.6	0.00006728	9.0	0.00006651
1.60	0.00006888	3.7	0.00006724	9.5	0.00006648
1.65	0.00006879	3.8	0.00006721	10.0	0.00006646
1.70	0.00006871	3.9	0.00006718		

INDEX.

Alinement (see pipelaying, tunnels)
Albuminoid ammonia 114
Arches:
   Abutments for, formula for calculating..................... 143
   Best form of.. 145
   Curve of pressure, calculation of.......................... 146-149
   Design, general rules governing............................ 149-153
   Failure, modes of... 145
   Joint of rupture of... 139
   Thickness, methods of calculating.......................... 137-143
Artesian wells, general discussion of........................... 74, 114
Axioms of hydraulics.. 9
Backfilling trenches for water pipe............................. 63
Back-pressure, loss of head from 57
Bends and elbows, loss of head by............................... 60
Calking:
   Directions for 66
   Lead, weight required....................................... 67
Center walls for earth dams..................................... 96
Chlorine 114
Coefficients, smoothness of pipes............................ 17, 18, 51
Concrete, composition for hydraulic work........................ 99
Conduits, masonry:
   Flow through, compared with pipe lines...................... 124
   Horse-shoe section, maximum flow of......................... 125
   Large sizes, form of section best for....................... 125
Core walls for earth dams....................................... 96
Croton basin, Rainfall available from........................... 77
Curve of pressure in arches..................................... 146-149
Dams:
   Character of, suitable to different conditions............... 82
   Concrete masonry work for................................... 99
   Foundation beds, determining suitability of................. 82
   Foundation pits, drainage of................................ 104
   Location, conditions governing.............................. 82
Dams, earth:
   Centre walls for.. 96
   Discharge outlets for....................................... 98, 122
   Embankments for 96, 98
   Spillways for 97
Dams, masonry:
   Base, construction of....................................... 95
   Classes of.. 82
   Design of high,
      Example illustrating.................................... 88-93, 122

INDEX.

Formulas for, unreliability of.. 93
Equilibrium of,
    Rectangular sections.. 82
    Trapezoidal sections.. 83
    Hydrostatic pressure of water against........................ 119
    Line of pressure, location of...................................... 86
    Plan, conditions governing selection of....................... 94
    Overturning, stability against................................. 82-86
Pressure on foundation masonry, safe amount.............. 87, 121
Sliding on base, stability against......................... 117-119
Stone masonry work for... 100
Stability, Vauban's principle.. 84
Diameter (see pipes, pipe lines, pipe line systems)
Discharge outlets for earthen dams........................... 98, 122
Discharge (see pipes, pipe lines, pipe line systems)
Drainage of foundation pits.. 104
Duty trials (see pumping engines)
Embankments for earth dams................................. 96, 98
Entry, resistance to
    Head, loss of due to.. 56
    Short horizontal pipes.. 12
Evaporation:
    From water surfaces... 116
    Sources, principal... 72
Fifth powers, Tables of... 48, 110
Fifth roots, calculation, use of logarithms in................. 28
Filters, mechanical,
    Action, method of... 128
    Cost of operation... 128
Filtration, general discussion of................................... 127
Fire service, quantity of water required for................. 116
Flow:
    Friction, loss of head from.. 56
    Horse shoe conduits, maximum through................... 125
    Pipe lines, branched,
        Branches variously placed, effect on............... 31-38
        Calculation, example illustrating.................. 28, 106
    Pipe lines,
        Calculation, example illustrating................... 19, 21
        Fundamental equations for............................ 18, 106
        Varying diameters, calculation of.......................... 22
    Pipe line system,
        Calculation, abbreviated method..................... 49, 107
        Calculation, formulas for.................................. 42, 106
        Communicating with two reservoirs....................... 51
    Through open channels.. 123
    Velocity of,
        Comparative through masonry conduits and smooth pipes.... 124

INDEX. 165

Grade of pipe lines, effect on................................. 14
Length of pipe effect on....................................... 12
Smoothness of pipe, effect on................................. 17
Short horizontal pipes... 11
Vertical pipes... 59
Formulas:
  Abutments for arches... 143
  Diameter of pipe carrying 100 gallons per capita in ten hours... 69
  Dimensioning spillways....................................... 81
  Duty of pumping engines...................................... 131
  Equilibrium of masonry dams.................................. 82
  Flow in pipe line system............................ 41, 42, 106
  Flow through long pipes................................. 18, 106
  Flow through open channels................................... 123
  Horse power for pumping................................. 69, 112
  Safe pressure on bottom courses of masonry dams......... 87, 121
  Spillways, approximate for determining....................... 81
  Storage reservoirs, capacity of.............................. 117
  Thickness of arch keys....................................... 138
  Velocity of falling bodies................................... 11
  Weight of cast iron pipe..................................... 168
  Weight of lead required for calking.......................... 67
Foundation blocks for water pipe............................... 65
Foundations:
  Dams, testing suitability of................................. 82
  Drainage of.. 104
Friction in pipes, head required to overcome.............. 12, 56
Grade (see pipes, pipe lines, pipe line systems, pipelaying, flow)
Head:
  Definition of
    Horizontal pipe lines...................................... 11
    Pipe lines with varying grades............................. 13
  Height of,
    Friction in pipes, amount required to overcome............. 12
    Resistance to entry, amount required to overcome........... 12
  Losses due to,
    Back pressure.. 57
    Bends and elbows... 60
    Changes in diameter of pipe 59
    Friction of flow... 56
    Resistance to entry.. 56
    Velocity, height required to produce....................... 56
  Loss of
    Definition of term... 55
    Hydraulic grade line, effect on............................ 63
    Pipe line systems, calculation of.......................... 56
    Practical importance of.................................... 63

INDEX.

Pressure in pipe lines due to.................................... 14
Heart walls for earth dams....................................... 96
Hydraulic grade line:
    Correct determination, importance of......................... 14
    Definition of term.. 9
    Determination for pipe lines with varying grades.............. 13
    Head, changes caused by loss of............................... 63
    Pipe lines, calculation for................................... 21
    Pressure in pipe lines, relation to........................... 25
    Varying steepness, effect on flow............................. 14
Hydraulic pressure, definition of term........................... 14
Hydraulics, axiomatic truths of.................................. 9
Hydrostatic pressure:
    Against plane surfaces.. 119
    Definition of term.. 14
Impurities, mineral, different kinds of.......................... 114
Joint of rupture, definition of.................................. 139
Line of pressure, in high masonry dams........................... 86
Masonry dams (see dams, masonry)
Masonry, stone, rules for constructing........................... 100
Mechanical filters (see filters, mechanical)
Mineral impurities in water...................................... 114
Nitrates and nitrites.. 114
Piezometric head:
    Branched pipe lines, calculation for...................... 23, 106
    Calculation of, example illustrating.......................... 21
    Pipe line system, calculation for......................... 49, 107
    Pipe line system, communicating with two reservoirs, calculation for.. 52
Piezometric height, definition of term........................... 10
Piezometric tubes, definition of................................. 14
Pipe, diameter of, having capacity of 100 gallons per capita in ten hours, formula for... 69
Pipelaying:
    Alinement, methods of... 65
    Backfilling trench.. 66
    Calking, directions for....................................... 66
    Calking, weight of lead required.............................. 67
    Foundation blocks for... 65
    General discussion.. 65
    Grades, method of preserving.................................. 65
Pipe lines:
    Diameter,
        Calculation, example illustrating......................... 20
        Changes in, loss of head from............................. 59
    Flow,
        Calculation of, example illustrating...................... 19
        Fundamental equations for............................. 18, 106
        Smoothness of pipe, effect on............................. 17

Varying diameter of pipe effect of............................ 22
Friction in, head required to overcome...................... 12, 56
Head of (see head)
Hydraulic grade line
    Calculation, example illustrating........................... 21
    Determination for varying grades........................... 13
    Length, measurement adapted for calculations................ 19
    Location with respect to hydraulic grade line................. 14
    Piezometric head, calculation of, example illustrating.......... 21
    Pressure in,
        General discussion...................................... 14
        Relation to hydraulic grade line......................... 25
    Uniform diameter equivalent to compound system, calculation of... 26, 27
Pipe lines, branched,
    Flow,
        Branches variously placed................................ 31-38
        Calculation of example illustrating...................... 28, 106
Pipe line system:
    Communicating with two reservoirs........................... 51
    Diameter of pipes for, calculation of......................... 39
    Flow,
        Calculation, abbreviated method........................ 49, 107
        Formulas for calculating........................ 42, 47, 106, 107
    Friction in... 12, 56
    Head, loss of, calculation of................................. 56
    Piezometric head, calculation, abbreviated method.......... 49, 107
Pipes:
    Cast iron, tables of weight and thickness..................... 68
    Friction in... 12, 56
    Long horizontal, velocity of flow in.......................... 12
    Short horizontal
        Discharge through, velocity of.......................... 11
        Resistance of entry to.................................. 12
    Smoothness, coefficients of................................. 17
    Vertical, flow through...................................... 59
Pressure (see pipes, pipe lines, pipe line systems)
Pumping horse power required, formula for.................. 69, 112
Pumping engines:
    Classes, description of different............................ 129
    Class, selection of, conditions governing..................... 129
    Duty of, formulas for...................................... 131
    Duty trials,
        Definition of... 130
        Methods of conducting................................ 130-137
    High duty, proper field of use............................... 130
    Types of, merits of different................................ 137
Pumps (see pumping engines)
Purity (see water supply)

Rainfall:
    Amount available from Croton Basin.......................... 77
    Methods of estimating.. 77
    Per square mile of drainage area............................. 78
Reservoirs, Storage:
    Capacity required, conditions governing...................... 78
    Spillways for, approximate formulas.......................... 81
Smoothness (see pipes, pipe lines, pipe line systems)
Spillways:
    Dams, earth.. 97
    Formulas, approximate, for................................... 81
Springs, quality of water supply from............................ 75
Storage, capacity required:
    Calculation of, example illustrating......................... 78
    Formula for calculating...................................... 117
Stone masonry, rules regarding construction..................... 100
Streams:
    Relative purity of large and small........................... 72
    Yield, method of estimatng................................... 77
Tables:
    Cast iron pipe, weight and thickness of...................... 68
    Coefficients of smoothness of pipes.......................... 18
    Explanation of... 154
    Fifth powers of pipe diameters............................... 48
Tubes, piezometric, definition of................................ 14
Tunnels:
    Alinement, possible accuracy of.............................. 126
    Construction of.. 125
Velocity (see flow, head)
Water supply:
    Artesian wells for, general discussion.................. 74, 114
    Evaporation, amount of.................................. 72, 116
    Filtration of.. 127
    Impurities, common mineral................................... 114
    Quality of:
        Chemical analysis inadequate to determine........... 71, 113
        General criterion of..................................... 71
        General discussion on................................. 71-76
        Large and small streams.................................. 72
        Springs, general purity of............................... 75
        Wells, impurity of....................................... 74
    Quantity required:
        Fire service... 116
        Future growth in population, provisions for.............. 76
        Per capita per 24 hours.................................. 76
        Sources, classification of............................... 72
        Storage for, example illustrating........................ 78
Wells:
    Artesian, general discussion............................ 74, 114
    Driven, quality of supply from............................... 74

www.ingramcontent.com/pod-product-compliance
Lightning Source LLC
Chambersburg PA
CBHW030247170426
43202CB00009B/659